Sound Intensity

Sound Intensity

Second edition

F.J. Fahy

Professor of Engineering Acoustics
Institute of Sound and Vibration Research
University of Southampton, UK

E & FN SPON
An Imprint of Chapman & Hall

London · Glasgow · Weinheim · New York · Tokyo · Melbourne · Madras

Chapman & Hall,
K

ondon SE1 8HN, UK

Blackie Academic & Professional, Wester Cleddens Road, Bishopbriggs, Glasgow G64 2NZ, UK

Chapman & Hall GmbH, Pappelallee 3, 69469 Weinheim, Germany

Chapman & Hall USA, 115 Fifth Avenue, New York, NY 10003, USA

Chapman & Hall Japan, ITP-Japan, Kyowa Building, 3F, 2-2-1 Hirakawacho, Chiyoda-ku, Tokyo 102, Japan

Chapman & Hall Australia, Thomas Nelson Australia, 102 Dodds Street, South Melbourne, Victoria 3205, Australia

Chapman & Hall India, R. Seshadri, 32 Second Main Road, CIT East, Madras 600 035, India

First edition 1989
Reprinted 1990
Second edition 1995

© 1989, 1995 F.J. Fahy

Typeset in 10/12pt Times by Thomson Press (India) Ltd, New Delhi
Printed in Great Britain by St Edmundsbury Press, Bury St Edmunds, Suffolk

ISBN 0 419 19810 5

A catalogue record for this book is available from the British Library

Library of Congress Catalog Card Number: 95–69215

∞ Printed on permanent acid-free text paper, manufactured in accordance with ANSI/NISO Z39.48-1992 and ANSI/NISO Z39.48-1984 (Permanence of Paper).

This book is dedicated to the memory
of my eldest son
Flying Officer Adrian Francis Fahy, RAF

Contents

Preface to the first edition

Prior to 1970, the possibility of routine measurement of the magnitude and direction of the flow of sound energy produced by noise sources in normal operational environments was just a pipe dream. Today, it is a reality; and the number of people planning, performing and interpreting such measurements is growing rapidly. In this monograph I attempt to give an account of the development of sound intensity measurement techniques up to the present day; to describe, in qualitative and quantitative terms, the energetics of sound fields; to explain the principles of transduction of sound intensity, and to describe and appraise the various forms of instrument and transducer probe; to summarise the state-of-the-art in the various areas of practical application of sound intensity measurement; and to explain the associated technical and economic benefits.

My principal purpose in writing this book has been to compile information about sound intensity and its measurement which is otherwise accessible only to those who have the time and energy to seek out and peruse the wide range of publications in which it appears. I have included, in Chapter 3, a brief exposition of the physical and mathematical foundations of the subject, with particular emphasis on those aspects of source mechanisms and kinematics which are not, in general, comprehensively covered in text books on the fundamentals of acoustics. Specialist acousticians may therefore prefer to start at Chapter 4 on 'Sound Energy and Sound Intensity'.

I have not addressed the subject of vibrational power flow in solid structures because I feel that, at the time of writing, this subject is still very much at the research stage, with as yet, relatively little evidence of practical utility; whereas the measurement of sound intensity in air has already become established through widespread use as a generally reliable procedure which offers substantial technical and economic advantages over pre-existing measurement methods. In terms of the timescale for writing and publishing a technical book, sound intensity measurement is an area of physical metrology which is still undergoing rapid development, and I beg the indulgence of the reader in regard to any obsolescent material.

My understanding of, and enthusiasm for, the subject of sound inensiy and its measurement have been greatly increased by my association during the past six years with the members of ISO/TC 43/SC1 Working Group 25 on 'The Determination of the Sound Power Levels of Sources Using Sound Intensity Measurement at Discete Points'. I wish to express my gratitude for their contribution of this book, albeit involuntary. The text has greatly benefitted from the comments and advice generously given by a number of colleagues who sacrificed valuable time and energy to read the original draft: in this regard I am deeply indebted to Elizabeth Lindqvist, J. Adin Mann III, Kazuharu Kuroiwa, Jiri Tichy and Hermann-Ole Bjor. I am also delighted to acknowledge the support of my wife Beryl, who never once mentioned how anti-social I had become while glued to my word processor, and the assistance of my youngest son Tom with the tedious task of proof reading. Originals of many of the figures were kindly made available by the authors of their source publications and I am indebted to James Mason for mastering the 'Mac' to produce a number of others. Finally, I wish to acknowledge the hospitality extended to my by Professor Claude Lesueur and his staff at INSA de Lyon, France, during my period of 3 months study leave in 1987, when a substantial part of the writing was done.

I hope that the book will be found to contain something of use to everyone concerned with the subject, whether beginner or expert.

F.J. FAHY

Preface to the second edition

Since 1987, when most of the first edition of *Sound Intensity* was written, major developments have taken place in the standardisation of procedures and instruments for the determination of sound powers of sources. National standards have been introduced in France, the Nordic countries, the United States and Canada, *inter alia*. The first International Standard, ISO 9614-1: 'Determination of sound power levels of sources using sound intensity— Part 1: Measurement at discrete points', has been published. As a result of many recent investigations into the effectiveness and accuracy of the scanning technique, examples of which are cited herein, a companion standard based upon the scanning technique has, at the time of writing, been brought to the draft standard, and should appear by 1995. IEC Standard 1043: 'Instruments for the measurement of sound intensity' was published in 1993.

As a consequence of critical comments directed at the organisation of the material of the first edition, the chapters on the principles and methods of application of sound intensity measurement have been integrated. New sections deal with transient and instantaneous sound intensity, and the section on complex intensity has been extended to consider more explicitly the relationship between energy transport and mean intensity distributions. A number of sound field indicators which have been proposed during the last five years are introduced and their significance is discussed. The section on instrumentation and calibration has been extended to incorporate a summary of the principal elements of IEC 1043, and further examples are presented of the application of sound intensity measurement to sound field characterisation and source location.

Readers have been very kind in bringing to my attention the errors in the first edition. I am much relieved to have the opportunity of correcting them in this edition.

I wish to express my appreciation of the assistance given to me during the preparation of this edition by Finn Jacobsen of the Technical University of Denmark, through frequent exchanges of correspondence. His publications have been a continuous source of inspiration to me and his constructive comments have been invaluable: the responsibility for the published material is, of course, entirely mine. I would also like to acknowledge the contribution to my knowledge and understanding of the subject made by my colleagues on ISO/TC43/SC1/Working Group 25 with whom I have worked over a period of more than ten years to develop the sound power standards.

As those readers who themselves have written a book will know only too well, it requires an enormous amount of labour to turn a manuscript into the finished product. I wish to acknowledge the skill which the editing and production team have brought to this task. My special thanks go to my dear wife Beryl for her loving support during the, sometimes fraught, process of writing this book, and for cheerfully performing the onerous task of correcting the proofs. Finally, I wish to express my gratitude to all those colleagues who have permitted me to include extracts from their publications.

Frank Fahy
Southampton
June 1995

CHAPTER ONE

Introduction

The physical phenomenon called 'sound' may be defined as a time-varying disturbance of the density of a fluid medium, which is associated with very small vibrational movements of the fluid particles. In this book, the frequency range of interest is assumed to be the so-called 'audio-frequency' range, which extends from about 20 Hz to 20 kHz. Audio-frequency vibrations can also occur in solid materials, such as steel or wood; these are always accompanied by sound in any fluid with which the solids are in contact. Such solid-borne vibrations may propagate in many different waveforms, unlike sound in fluids. This phenomenon is termed 'structure-borne sound', derived from the more concise German word *Körperschall*.

Sound in a fluid depends for its existence upon two properties of the medium: (i) the generation of pressure in response to a change in the volume available to a fixed mass of fluid, i.e. change of density; (ii) the possession of inertia, i.e. that property of matter which resists attempts to change its momentum. Both the forces generated by volumetric strain of fluid elements, and the accelerations of those elements, are related to their displacements from positions of equilibrium. The resulting interplay produces the phenomenon of wave motion, whereby disturbances are propagated throughout the fluid, often to very large distances. The nature of sound, and the behaviour of sound waves, are the subjects of Chapter 3, in which emphasis is placed on the kinematic features of sound fields, which are illustrated by various simple examples.

Sound waves in fluids involve local changes (generally small) in the pressure, density and temperature of the media, together with motion of the fluid elements. Fluid elements in motion have speed, and therefore possess kinetic energy. In regions where the density increases above its equilibrium value, the pressure also increases; consequently, energy is stored in these regions, just as it is in a compressed spring. This form of energy is termed potential energy. Textbooks often introduce the subject of sound in terms of simple harmonic motion, which may lead students to believe, like Isaac Newton, that such motion is natural to fluid particles disturbed from equilibrium. In fact, fluid particles will oscillate continuously only if waves are continuously generated by a source, or if, once generated, they repeatedly retraverse a fluid region via reflections from surrounding boundaries. It is not intuitively obvious that in either case energy will be transported from one location to another; it seems much more likely that it will just be transferred to and fro between adjacent fluid elements. Consider, therefore, a transient sound created in the open air, for example by a handclap. A thin shell of disturbance will spread out all around the source, travelling at the speed of sound. Within this disturbed region the fluid particles will be temporarily displaced from their equilibrium positions, and the pressure, density and temperature will temporarily vary from their equilibrium values. Once the disturbance has passed, everything is just as it was before—the fluid particles are once more at rest in their original positions and do not continue to oscillate.

It is quite clear from this qualitative description of wave propagation that the potential and kinetic energies created by the action of the source on the air immediately surrounding it are transported with the disturbance; they cannot disappear, except through the action of fluid friction (viscosity), and other dissipative processes, which are known to have rather small effect at audio frequencies. *Sound Intensity* is a measure of the rate of transport of the sum of these energies through a fluid: it is more explicitly termed *Sound Power Flux Density*. Chapter 4, which deals with the energetics of sound fields, shows that sound intensity is a vector quantity equal to the product of the sound pressure and the associated fluid particle velocity vector.

What, may be asked, is the importance of being able to measure sound intensity, rather than sound pressure, which we have been able to measure with increasing accuracy every since Wente devised the condenser microphone over 60 years ago? The short answer is that, by means of the application of principles explained in Chapters 5 and 9, the user of intensity measurement equipment may quantify the sound power generated by any one of a number of source systems operating simultaneously within a region of fluid, thereby greatly assisting the efficient and precise targeting and application of noise control measures.

The first patent for a device for the measurement of sound energy flux was granted to Harry Olson of the RCA company in America in 1932 (incidentally, also an *annus mirabilis* in particle physics). The first commercial sound intensity measurement systems were put on the market in the early 1980s. Why the 50-year delay? Therein hangs a long tale, the details of which will be related in Chapter 2. It may be basically attributed to the technical difficulty of devising a suitably stable, linear, wide frequency band transducer for the accurate conversion of fluid particle velocity into an analogue electrical signal, together with the problem of producing audio-frequency electrical filter sets having virtually identical phase responses. As explained in Chapter 6, current instruments are of three basic types: signals from the transducers are passed either through digital or analogue filters, followed by sum and difference circuits and integrators, or processed using discrete Fourier transform algorithms to produce the cross-spectral estimates to which sound intensity is proportional. Currently available probes comprise either two nominally identical pressure-sensitive transducers, or one pressure transducer in combination with an ultrasonic particle velocity transducer.

All measurements are subject to error. Errors associated with the specific form of transducer system used to provide the signals from which intensity estimates are obtained are explained and quantified in Chapter 5, and those generated by the measurement system are treated in Chapters 6 and 7. The latter also presents a comprehensive analysis of random errors associated with the signal processing procedures employed. The nature of sound fields near sources and the sound field indicators used to classify sound fields are described in Chapter 8.

The important advantages which accrue from the availability of accurate sound intensity measurement systems are consequent upon the fact that sound intensity is a vector quantity, having both magnitude and direction, whereas pressure is a scalar quantity, possessing only magnitude. As explained in Chapter 9, this extra dimension allows the contribution of one steady sound source operating among many to be quantified under normal operating conditions in the operational environment: the need for special-purpose test facilities is largely obviated, a development which confers clear and substantial economic benefits. The major application of sound intensity measurement is to the determination of the sound power output from individual sources in the presence of others. The advantage of *in situ* measurement of sound power extends beyond the obviation of the need to construct or hire expensive, special-purpose test facilities. Many sources of practical importance are either too big, too heavy, too dangerous, or too dependent upon ancillary equipment to be transported to a remote test site. In addition, the *in situ* application of approximate methods based on the measurement of sound pressure requires microphones to be located at a distance from the source comparable with its maximum dimension, which, in many cases, puts the microphone into regions where noise from interfering nearby sources is comparable with, or exceeds, that of the source under investigation: by contrast, valid sound intensity measurements may be made at any distance from a source. A particular advantage of sound intensity measurement to manufacturers is that production test bays may serve to double as sound power check facilities. Sound intensity measurement may additionally be used to investigate the distribution of sound power radiation from any source, including vibrating partitions which separate adjacent spaces, so that the various regions may be placed in rank order; this has been found to be of particular utility in the case of automotive engine noise, and in detecting weak regions in partitions.

Other applications of sound intensity measurement which are at present (1994) less well developed than sound power determination are the *in situ* evaluation of the acoustic impedance and sound absorption properties of materials, and the *in situ* determination, under operational conditions, of the sound power generated by fans, together with the performance of associated in-duct attenuators. The latter application forms the subject of Chapter 11.

The advent of practical, reliable, sound intensity measurement may be seen as one of the most important developments in acoustic technology since the introduction of digital signal processing systems. It is of great value to equipment designer, manufacturer, supplier and user, and, in particular, to the specialist acoustical engineer concerned with the control and reduction of noise.

CHAPTER TWO

A brief history of the development of sound intensity measurement

In the nineteenth century, Rayleigh devised a suspended disc system of which the deflection is proportional to the square of particle speed. This may be said to indicate intensity, but, of course, it only does so in simple travelling wave fields: it is not, however, a practical measuring device, because it is also disturbed by air movement. The recorded history of attempts to measure the flow of sound energy in complex sound fields goes back over 60 years, to 1931 when Harry Olson of the Radio Corporation of America submitted an application for a patent for a 'System responsive to the energy flow of sound waves'. The patent was granted in the following year (Figs 2.1 and 2.2). As explained by Wolff and Massa,[1] the 'Field Wattmeter' system was designed to process the signals from a pressure microphone and a particle velocity microphone using the 'quarter square' multiplication principle by which the product of the two signals was obtained from the difference between the squares of the sum and difference of the signals. In an article published many years later,[2] Olson describes a suitable microphone (Fig. 2.3), and a development of the basic wattmeter to incorporate band pass filters (Fig. 2.4). Curiously, the literature appears to contain no evidence that this device achieved significant practical utilisation.

During the following 20 years, sporadic attempts were made to develop measurement systems with reasonable frequency ranges, but serious problems were encountered with performance instability, and excessive sensitivity to ambient conditions, such as wind, humidity and temperature. In the early 1940s, Enns and Firestone presented one of the very few theoretical analyses of sound energy flux in source fields to appear before 1960 (Fig. 2.5).[3] At the same time, they, together with Clapp, used a combination of a ribbon velocity microphone and two crystal pressure microphones to investigate sound intensity fields in a standing wave tube and in a reverberant room.[4] The system performed well in the former situation, but poorly in the latter (which, with hindsight, does not surprise us today). In introducing a two-pressure-microphone technique for the purpose of *in situ* determination of the acoustic impedance of material samples in 1943, Bolt and Petrauskas[5] paved the way

Dec. 27, 1932. H. F. OLSON 1,892,644

SYSTEM RESPONSIVE TO THE ENERGY FLOW OF SOUND WAVES

Filed May 29, 1931

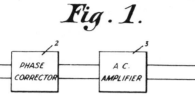

Fig. 1.

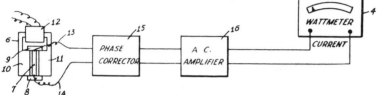

Fig. 2.

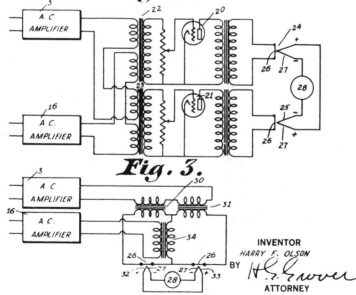

INVENTOR
HARRY F. OLSON
BY
ATTORNEY

Fig. 2.1. Olson's patent circuits for a 'System responsive to the energy flow of sound waves' (United States Department of Commerce: Patent and Trademark Office).

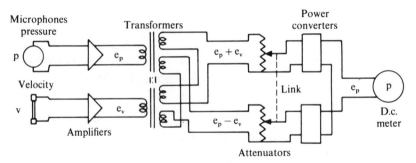

Fig. 2.2. Schematic diagram of Olson's field-type wattmeter.[2]

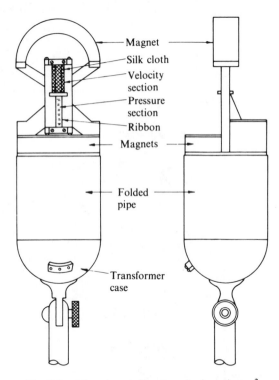

Fig. 2.3. Olson's uni-directional microphone.[2]

for development, in the distant future, of a range of two-microphone measurement methods. Twelve years later, Baker[6] reported his attempts to use a hot wire anemometer in combination with a pressure microphone to measure sound intensity; unfortunately, the system was far too sensitive to extraneous air movements to be suitable for field measurement applications.

In 1956, Schultz[7] made a major contribution to the development of practical sound intensity measurement systems during research for his PhD.

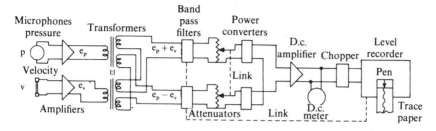

Fig. 2.4. Schematic diagram of a field-type acoustic wattmeter with band-pass filters and level recorder.[2]

He implemented the principle employed by Bolt and Petrauskas, which is widely used today, by which a particle velocity signal can be obtained by integrating the difference between signals produced by two small pressure-sensitive transducers spaced a small distance apart in terms of the wavelength of sound at the highest frequency of interest. Unfortunately for the development of practical measurement systems, Schultz's disc-like transducers were placed back-to-back, with their surfaces a very small distance apart. This configuration placed extreme demands on the electronic circuitry of the day, and, although he demonstrated satisfactory performance under laboratory conditions in relatively simple sound fields, attempts to survey the sound field generated by a sound source in a rigid-walled enclosure were disappointing. As Schultz explained later,[8] it was not only the inadequate performance of the measurement system under highly reactive conditions which produced the problem; lack of a comprehensive theoretical analysis, and therefore of physical understanding, of the enclosed sound field, also contributed to the resulting lack of confidence in the measured results.

It has always appeared to me something of an irony that all the theoretical tools for the analysis of sound intensity fields had been available for many years, and indeed very complex sound pressure fields had been analysed by means of elegant mathematical techniques (e.g. Stenzel's analysis of piston near fields),[9] but there seemed to be little interest in investigating the associated intensity fields. Perhaps Morse's prescient remark in his excellent book *Vibration and Sound*,[10] 'Unfortunately (or fortunately, perhaps) we seldom measure sound intensity', was symptomatic of the prevailing general attitude to this second-order acoustic quantity. Schultz[8] offers an explanation of the 'lean years': he says 'I have also wondered why, over the past 40 years, a number of researchers have developed acoustic wattmeters and have published a few test results to show that the device really works, and then nothing more is heard of the matter. The reason for this has recently begun to dawn on me: I suspect each man went on to make a few more measurements with his new wattmeter and then just kept quiet because he was appalled at the inexplicable results!'

We can learn a useful lesson from these early experiences: attempts to interpret measured distributions of sound intensity without having a thorough understanding of the nature of sound energy flux are, indeed, very dangerous.

In the late 1960s, Mechel,[11] Odin[12] and Kurze[13] demonstrated the relationships between the active and reactive components of sound intensity and, respectively, the spatial gradients of phase and squared pressure. These relationships are today effectively implemented in the indirect spectral technique of intensity measurement.

Pioneering contributions in the area of application of sound intensity measurement to the determination of the sound power radiated by complex sources, were made in the early 1970s by research workers in South Africa; van Zyl, Anderson and Burger, among others.[14,15] Although they used a combination of pressure and velocity microphones in their early work, they later realised the superiority of the combination of two nominally identical pressure microphones, and developed the first analogue intensity meter to have a wide frequency range and large dynamic range.[16] This instrument was developed from a prototype constructed by van Zyl for a Master's thesis in 1974.[17] Subsequently, this group developed a range of intensity meters of increasingly high performance as a small commercial venture. In the second half of the decade, a low frequency (50–500 Hz) analogue intensity meter was developed in Switzerland, by Lambrich and Stahel[18] for investigations inside cars, and a hybrid analogue–digital device was developed by Pavic[19] in Yugoslavia. In 1977, the author published a technique for measuring sound intensity using two condenser microphones and one sound level meter, based upon Olson's signal processing principle: estimates were also presented of the bias errors incurred in plane wave, monopole and dipole fields by the use of the finite difference approximation inherent in the two-microphone method.[20] The ISVR analogue sound intensity meter which developed out of this work largely solved the outstanding problem of the phase matching of filter sets by use of the newly available charge transfer devices.

Two commercially developed analogue sound intensity measurement systems eventually became available. The Metravib instrument[21] employed a probe incorporating three inexpensive miniature electret microphones, with switching circuitry to optimise the microphone separation distance for low and high frequency ranges. The Bruel & Kjaer instrument[22] employs high quality condenser microphones in face-to-face arrangement; the filters and integrator are based upon analogue devices, and the multiplication and subsequent processes employ digital circuitry.

During the 1970s, digital signal processing technology developed rapidly, and came into widespread use in the form of stand-alone FFT analysers. At last, reliable and fast phase measurements became possible at the touch of a key; the implications for intensity measurement were profound. In 1975, I was preparing a proposal for an undergraduate project on sound intensity measurement. On writing down the time domain form of the intensity in terms of two microphone signals, and converting to the frequency domain, it became clear that it was only necessary to generate the imaginary part of the cross spectrum to obtain the intensity. The remarkable implication was that anyone who possessed two high quality microphones, together with an FFT analyser,

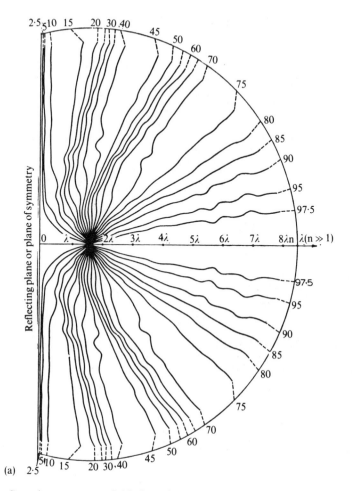

Fig. 2.5. Sound power vector field of a point source: (a) near a totally reflecting rigid plane; (b) near a totally reflecting free plane.[3]

effectively possessed a sound intensity meter. Miller, in his undergraduate project,[23] was possibly the first person to apply this result, although, as indicated below, it subsequently transpired that others were treading the same path.

The Year of Our Lord 1977 proved to be a vintage year in the development (or at least in the revelation) of digital signal processing techniques for the measurement of sound intensity. Alfredson,[24] in Australia, employed two face-to-face condenser microphones to evaluate the sound intensity radiated by a multi-cylinder engine by evaluating the Fourier coefficients of the engine harmonics, and expressing the tonal intensity in terms of the real and imaginary parts. He did not, however, separate the microphones by a solid cylindrical plug which was later found necessary for the definition of acoustic separation, and the suppression of undesirable diffraction effects. His

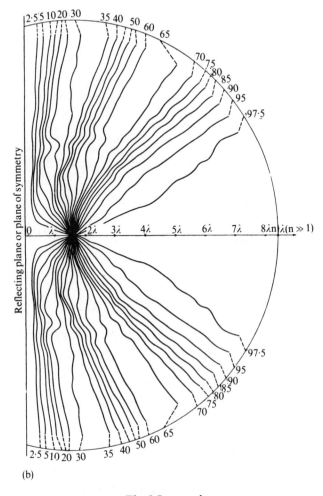

(b)

Fig. 2.5—*contd.*

theoretical expression was the tonal equivalent of the cross-spectral expression used by Miller.

In France, Lambert and Badie-Cassagnet[25] derived an expression for intensity based upon the complex Fourier transforms of signals from two closely spaced microphones, which they evaluated on a digital computer. They tested their measurement system on a reference sound source operating in an anechoic chamber (near field and far field measurements) and in a highly reverberant enclosure. They also investigated the influence of a parasitic source in a semi-anechoic room. They obtained consistent results in a frequency range of 63 Hz–10 kHz, except below 200 Hz in the reverberant enclosure, where errors of up to 6 dB were attributed to imperfect matching of the transducer channels.

In the United States of America, Chung, Pope and colleagues developed an intensity expression based upon the imaginary part of the cross-spectral density between two closely spaced pressure microphones, which is the spectral equivalent of the expressions of Alfredson and Lambert, and applied it to the measurement of the sound radiation of automotive components.[26] Chung later introduced a practical technique for correcting phase mismatch between microphone channels.[27]

In 1977, I also published the cross-spectral formulation used by Miller in his student project,[28,29] which is essentially the same as that of Chung, presenting results of measurements in a standing wave tube and on an automotive engine. The scene was set for an explosion of applications of the new measurement techniques.

In the 1970s, the combination of pressure and particle velocity microphones was discarded in favour of a combination of two nominally identical pressure microphones. However, more recently, a new form of velocity microphone based on the principle of convective Doppler shift of an ultrasonic beam has been developed.[30,31]

The previous decade has seen introduced a number of commercial sound intensity measurement systems; some are based on FFT analysis, and others employ digital filters. Manufacturers have made a vital contribution to the practical implementation of the principles of sound intensity measurement by developing optimised configurations of transducer probes, and assisting in the development of an instrument standard.

Research has improved survey procedures for the determination of the sound power of one source operating in the presence of others and has led to the development and publication of national and international measurement standards (see Appendix). Current research concerns, among other things, applications in duct and building acoustics; power flux line mapping; and imaging of acoustic sources by near field microphone array techniques, which generate intensity distributions as a by-product.

Readers may be disappointed to find no mention of 'structural intensity' in this monograph. The measurement of vibrational energy flux in solid structures is considerably more difficult and complicated than the measurement of sound intensity in fluids. Consequently it is still in the relatively early stages of research and development, and has yet to become established as a routine measurement tool: it will no doubt form the subject of a future monograph in its own right.

CHAPTER THREE

Sound and sound fields

3.1 THE PHENOMENON OF SOUND IN FLUIDS

The physical phenomenon known as 'sound' in a fluid (i.e. a gas or a liquid) essentially involves time-varying disturbances of the density of the medium from its equilibrium value: these changes of density are in most cases extremely small compared with the equilibrium density (typically of the order of 10^{-7}–10^{-5}). They may be attributed to changes in the volume of space occupied by a given mass of fluid, changes of shape not being of consequence in the case of sound waves: it is the volumetric strain, or dilatation, undergone by an elemental mass of fluid which matters. The static elastic nature of air in its response to volumetric strain is easily demonstrated by closing the outlet hole in a bicycle pump with a finger and depressing, and then releasing, the plunger; the plunger returns almost to its original position on release. It is not so easy to find an everyday phenomenon to demonstrate that liquids, such as water, are also elastic; this is because, in response to a given volumetric strain, they generate much larger internal stresses, and hence, reaction forces. Surprisingly, it took little longer to establish the 'compressibility', and hence sound speed, of water than that of air. The Swiss physicist Daniel Colladon measured its value in 1826 in response to the offer of a prize by the Paris Academy of Science, and the result was validated by a measurement of the speed of sound in Lake Geneva by Colladon and his colleague, the mathematician Charles Sturm, in the same year. The 'correct' speed of sound in air was evaluated theoretically by Pierre Simon Laplace in 1816, and the ratio of specific heats was accurately determined in 1819, a century or more after Isaac Newton and Leonhard Euler, among others, had presented incorrect theoretical predictions of the sound speed by assuming an isothermal process.

It must be clearly understood that fluid elasticity is not analogous to that observed in a bouncy rubber ball. Rubber is not acoustically dissimilar to water, in that it is almost incompressible: the bounciness derives not from changes of volume, but from changes of shape; the reaction forces are caused by shear distortion. The same applies to rubber mats which are used as vibration isolators; they are only effective if they are allowed to bulge laterally. (This is why it is quite misleading to try to assess the likely effectiveness of a sheet of resilient material by squeezing it between finger and thumb.)

Since normal fluids are homogeneous and isotropic, possessing the same properties in all directions (unlike, for example, woven materials), the effect of an *isolated, localised* density disturbance spreads out uniformly in all directions in the form of a wave. The phenomenon of 'directivity', which is exhibited by all *spatially extended* sources, such as a loudspeaker cone, is the result of interference between the elemental fields emanating from the various parts of the source region which lie at differing distances from any observation point. The phenomenon of 'diffraction', in which sound 'bends' round solid obstacles in its path, may be understood qualitatively in terms of Huygens' principle, which states that every point on a wavefront may be considered to act like a point source of sound. For example, if a spherical wavefront is replaced by a uniform set of point sources distributed uniformly over the wavefront surface, the *outgoing* interference field produced by the superposition of all the elemental spherical fields is readily seen to take the form of a further spherical wavefront. 'Removal' of portions of the wavefront by an obstruction may be seen, qualitatively, to lead to distortion of the shape of the propagating wavefronts, and to the formation of incomplete sound 'shadows'. Comprehension of the phenomena of interference and diffraction is vital to the proper understanding of sound intensity fields.

3.2   THE RELATIONSHIP BETWEEN PRESSURE AND DENSITY IN GASES

In the phenomenon of solid elasticity, stress resulting from distortion is, within a finite range of linear behaviour, proportional to the material strain. Most acoustic disturbances are so small that linearity is closely approximated; acoustic non-linearity in air becomes significant at sound pressure levels exceeding about 135 dB. Elastic stresses can take two forms, namely, normal and shear. In fluids, the latter can only be generated by relative motion, and, by definition, cannot be sustained statically. It is the normal fluid stresses, or pressures, which are of primary importance in the mechanism of sound propagation. The coefficient relating small pressure changes to small volumetric (dilatational) strains of a fluid is termed the 'bulk modulus'. The value of the bulk modulus of a gas depends upon the type of gas, and the conditions of the volumetric change. Contrary to Newton's assumption, sound in gas is not an isothermal process, in which the temperature remains constant, but very nearly an adiabatic phenomenon, in which no significant heat exchange occurs between instantaneously hotter and cooler regions.

An expression for the adiabatic bulk modulus of a gas may be derived by consideration of a graph of the variation of pressure with density (Fig. 3.1): P is the fluid pressure and P_0 is the equilibrium (mean, static) pressure; ρ is the density and ρ_0 is the mean density. The exponent γ is the ratio of specific heats of constant pressure and constant volume.

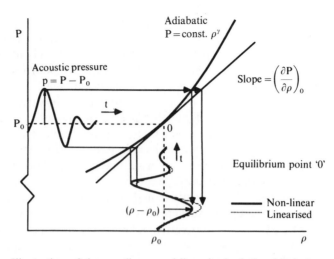

Fig. 3.1. Illustration of the non-linear and linearised relationship between acoustic pressure and density.

The adiabatic form of this graph applies to any such changes undergone by a fluid, but, in the case of acoustic disturbances, the difference between the instantaneous pressure P and the mean pressure P_0 is termed the acoustic pressure, symbolised by p. Figure 3.1 illustrates how P, and hence p, vary with ρ. It is seen that the relationship is non-linear for large deviations from equilibrium: however, it is asymptotically linear as the fractional change in pressure p/P_0 tends to zero. To give some idea of the order of magnitude of this ratio we note that for a sound pressure level of 100 dB it is 2×10^{-5}. The fractional change of density $(\rho - \rho_0)/\rho_0$, also known as the condensation s, is of the same order.

For small changes, the relationship between acoustic pressure and density change is, to first-order approximation (Taylor series expansion)

$$p = \delta P = (\partial P/\partial \rho)_0 \delta \rho \qquad (3.1)$$

where the subscript 0 means 'evaluated at equilibrium'. This is not directly a relationship between stress and strain, and so the term $(\partial P/\partial \rho)_0$ is not the bulk modulus. The relationship between density change and volumetric strain may be derived by considering a fixed mass of gas of which the volume is changed by a small proportion.

The statement of conservation of mass of an element of equilibrium volume V_0 is

$$\rho V = \rho_0 V_0 \qquad (3.2)$$

and therefore the volumetric strain is

$$\delta V/V = - \delta \rho/\rho \qquad (3.3)$$

Hence, (3.1) may be written, to first order,

$$p = -\rho_0(\partial P/\partial\rho)_0(\delta V/V) \tag{3.4}$$

which gives the bulk modulus as $\rho_0(\partial P/\partial\rho)_0$. For adiabatic changes, P/ρ^γ is constant; hence, $\rho_0(\partial P/\partial\rho)_0 = \gamma P_0$, and

$$p = \gamma P_0(\rho - \rho_0)/\rho_0 = \gamma P_0 s \tag{3.5}$$

Later we shall find that $\gamma P_0 = \rho_0 c^2$, where c is the speed of propagation of sound, so that the ratio of pressure fluctuation p to density fluctuation $(\rho - \rho_0)$ is equal to the square of the speed of sound in all regions of a sound field where the linearisation approximations hold good.

The bulk modulus of a liquid is not so easily derived from first principles, and depends upon a number of variables including mean pressure, adulteration by other materials, and mean temperature, *inter alia*. This volume is primarily concerned with sound in air, and therefore only the approximate expression for the speed of sound in water is given:[32]

$$c = 1402 \cdot 7 + 488t - 482t^2 + 135t^3 + (15 \cdot 9 + 2 \cdot 8t + 2 \cdot 4t^2)(P_g/100) \tag{3.6}$$

where P_g is the gauge pressure in bar and $t = T/100$, with T in °C.

3.3 DISPLACEMENT, VELOCITY AND ACCELERATION OF FLUID PARTICLES

In order to relate the forces and motions involved in fluid dynamic processes it is necessary to refer the positions of fluid particles to a frame of reference in which Newton's second law of motion may validly be applied. For the purposes of analysis of sound fields, a frame of reference fixed in the surface of the earth is adequate.

The position of a particle is described by its position vector **r**, as shown in Fig. 3.2. Particle displacement is symbolised by vector ξ. The particle velocity **u** is defined as $\partial\xi/\partial t$. In fluid flow it is not quite so straightforward to derive an expression for particle acceleration as it is in solid mechanics because the particles may mix in very complicated flow patterns. The velocity of a particle may vary both in space (x, y, z) and in time (t). Thus, a particle in a steady flow may change its velocity by virtue of a change in position in space, i.e. by moving into a region where the flow velocity is different from that in its former position. In an unsteady flow, the velocity of particles at any fixed position in space varies by virtue of the progress of time. Hence a small change of particle velocity may be expressed as

$$\delta\mathbf{u} = (\partial\mathbf{u}/\partial t)\delta t + [(\partial\mathbf{u}/\partial x)(\partial x/\partial t) + (\partial\mathbf{u}/\partial y)(\partial y/\partial t) + (\partial\mathbf{u}/\partial z)(\partial z/\partial t)]\delta t \tag{3.7}$$

Therefore the total acceleration is

$$\mathbf{Du}/\mathbf{D}t = \partial\mathbf{u}/\partial t + u(\partial\mathbf{u}/\partial x) + v(\partial\mathbf{u}/\partial y) + w(\partial\mathbf{u}/\partial z) \tag{3.8}$$

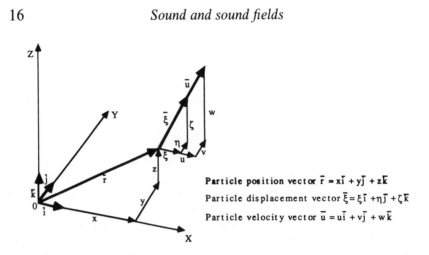

Fig. 3.2. Vector representation of particle position, displacement and velocity.

where u, v and w are the Cartesian components of the particle velocity vector **u**.

The first term in this expression represents acceleration of fluid particles due to unsteadiness (time variation) of the fluid motion at any point fixed in space: the other three constitute the 'convective derivatives', so-called because they express acceleration due to convection of a particle from one point in a flow to another point at which the velocity vector is different. This convective component is non-zero in all but a totally uniform steady flow. In all except the strongest sound fields, for example in the exhaust ducts of internal combustion engines, or in regions of mean fluid flow, as in ventilation ducts, the ratio of the magnitudes of the convective derivatives to the 'unsteadiness' derivative is of the order of the condensation s, which we have already seen to be typically of the order of 10^{-5}. We shall assume henceforth, except in Chapter 11 on sound intensity in flow, that

$$\mathbf{D}\mathbf{u}/\mathbf{D}t = \partial \mathbf{u}/\partial t \qquad (3.9)$$

It is not valid to make this approximation in regions of turbulent flow in which the products of particle velocities and their spatial derivatives are non-negligible. In fact, the sound generation mechanism of unsteady flows can be partially explained in terms of the momentum fluctuations represented by these terms.

3.4   EQUATION OF CONSERVATION OF MASS

The relationship between density and volumetric strain, previously derived in eqn (3.3), may also be derived using the Cartesian co-ordinate axes system used in the previous section and in Fig. 3.2.

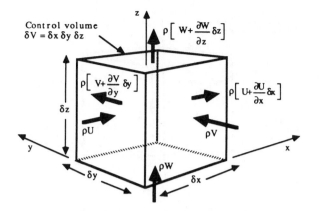

Fig. 3.3. Mass flux through a fluid control volume.

In Fig. 3.3 the rate of mass flow into the control volume is seen to be

$$(\rho u)\delta y \delta z + (\rho v)\delta x \delta z + (\rho w)\delta x \delta y$$

and the rate of mass flow out is

$$(\rho u + (\partial(\rho u)/\partial x)\delta x)\delta y \delta z + (\rho v + (\partial(\rho v)/\partial y)\delta y)\delta x \delta z + (\rho w + (\partial(\rho w)/\partial z)\delta z)\delta x \delta y$$

$$(3.10)$$

The net mass outflow must be balanced by a decrease in the density of the volume; thus

$$[\partial(\rho u)/\partial x + \partial(\rho v)/\partial y + \partial(\rho w)/\partial z]\delta V = -(\partial \rho / \partial t)\delta V$$

where $\delta x \delta y \delta z = \delta V$.

Order of magnitude analysis shows that this non-linear equation may be linearised in the case of small disturbances[32] to yield

$$\rho_0(\partial u/\partial x + \partial v/\partial y + \partial w/\partial z) + \partial \rho/\partial t = 0 \qquad (3.11)$$

Integration of this equation with respect to time yields the equivalent of eqn (3.3), showing that volumetric strain in Cartesian co-ordinates is

$$\delta V/V = \partial \xi/\partial x + \partial \eta/\partial y + \partial \zeta/\partial z$$

where ξ, η, ζ are the Cartesian components of the particle displacement vector ξ. This expression is known mathematically as the 'divergence' of the particle displacement, seen here to be an apt term.

3.5 FLUID MOMENTUM EQUATION

Fluid particles, like any other matter, obey Newton's laws of motion to a degree dependent upon the suitability of the chosen frame of reference. The

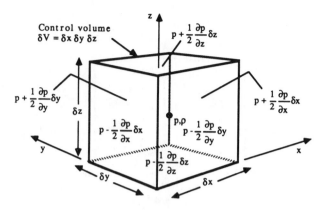

Fig. 3.4. Pressure gradients in a fluid.

mass of fluid contained within the control volume shown in Fig. 3.4, is subject to differences of fluid pressure on opposing parallel faces because of the existence of spatial gradients of pressure in the fluid. Application of the *linearised* expression for particle acceleration (eqn (3.9)), yields the following momentum equations in the three co-ordinate directions:

$$\partial p/\partial x = -\rho_0 \partial u/\partial t \qquad (3.12a)$$

$$\partial p/\partial y = -\rho_0 \partial v/\partial t \qquad (3.12b)$$

$$\partial p/\partial z = -\rho_0 \partial w/\partial t \qquad (3.12c)$$

(It must be noted carefully that the linearisation procedures employed in the foregoing sections lead to the eventual conclusion that sound cannot be generated by turbulent mixing processes in the absence of solid surfaces, such as those which occur in jet effluxes—a paradox not resolved until 1952 by Lighthill: a prime example of throwing the baby out with the bath water!)

3.6 THE WAVE EQUATION

Relationships between variations of physical quantities in time and space are inherent to the propagation of disturbances by wave phenomena. The mathematical expression of these relationships is called the 'Wave Equation', the form of which depends upon the nature of the wave motion, the nature of the wave-bearing medium and the co-ordinate system employed in its derivation. So far, we have derived relationships between acoustic pressure, fluid density, and the kinematic variables particle displacement, velocity and acceleration. These are clearly not independent of each other, and it is possible by means of simple mathematical manipulation of the above equations to derive equations in only one dependent variable.[32] The most commonly used form is that in terms of acoustic pressure,

$$\partial^2 p/\partial x^2 + \partial^2 p/\partial y^2 + \partial^2 p/\partial z^2 - (1/c^2)\partial^2 p/\partial t^2 = 0 \qquad (3.13a)$$

in which $c^2 = \gamma P_0/\rho_0$. In the special case of harmonic time dependence, this becomes the Helmholtz equation. For frequency ω

$$\partial^2 p/\partial x^2 + \partial^2 p/\partial y^2 + \partial^2 p/\partial z^2 + k^2 p = 0 \qquad (3.13b)$$

where $k = \omega/c$.

In mathematical terminology it is said to be the 'homogeneous' form of the acoustic scalar wave equation, because it contains no forcing term, or input, on the right-hand side. It governs the behaviour of sound in any region of fluid in which the assumptions made in its derivation are satisfied. Because we have not included in our analysis any boundary conditions on the fluid, or any active sources of sound, solutions of this equation answer the question 'What form of waves *can* exist in the medium?', but not the question 'What specific wavefield *does* exist in the medium?'. The particular field created by any specific form of sound source, in any specific region of fluid, can only be predicted theoretically once sources and boundary conditions are defined. In fact, eqn (3.13) is only one form of the wave equation, particular to the rectangular Cartesian co-ordinate system: the form, *but not the physical meaning*, varies with the chosen co-ordinate system, as shown in textbooks (e.g. Ref. 32). The linear relationship eqn (3.5) clearly shows that acoustic pressure may be replaced by density perturbation in eqn (3.13) and, indeed, it is also satisfied by fluid temperature.

3.6.1 The inhomogeneous wave equation

The homogeneous wave equation (3.13) describes the acoustic behaviour of an ideal fluid in which arbitrary small disturbances from equilibrium exist. Physical sources of sound which cause such disturbances are extremely diverse in their forms and characteristics. However, the basic sound-generating mechanisms of all sources may be classified into three categories:

(i) fluctuating volume (or mass) sources;
(ii) fluctuating force sources;
(iii) fluctuating excess momentum sources.

If, in a region of a volume of fluid, there operates a mechanism by which fluid volume is actively displaced by an independent agent in an unsteady fashion, a new term must be introduced into the linearised mass conservation equation (3.11);

$$\rho_0(\partial u/\partial x + \partial v/\partial y + \partial w/\partial z) - \rho_0 q = -\partial \rho/\partial t \qquad (3.14)$$

where q is the rate of displacement, or introduction, of fluid volume (the volumetric velocity) per unit volume: $\rho_0 q$ represents a rate of mass introduction per unit volume. A good example of such a source is the acoustic siren, in which 'puffs' of air are forced into an otherwise quiescent atmosphere through cyclically operating valves. The air in the vicinity of each outlet hole finds itself periodically displaced by the ejected air, and reacted against by the inertia of the immediately surrounding air, with the consequent time-dependent variation of fluid density, and the generation of a propagating sound field. In deriving the wave equation (3.13), the mass conservation equation is differentiated with respect to time. Hence the term which appears on the right-hand side of the inhomogeneous wave equation corresponding to this forced volume displacement term is $\rho_0\, \partial q/\partial t$, indicating that it is a rate of change of a rate of volume displacement (volumetric acceleration) which gives rise to sound. Obviously, a *steady* stream of fluid issuing from a pipe does not generate noise in the same way as a siren, although it can generate noise by the process of turbulent mixing, which is a non-linear, category (iii) mechanism, many orders of magnitude less efficient than that of volume displacement. A very common example of the first category of source is the vibrating, impervious surface. Fluid adjacent to the surface is displaced in a time-dependent fashion, and sound is consequently radiated. The effectiveness with which any vibrating surface radiates sound energy depends upon both the magnitude of the surface acceleration, and the form of its distribution over that surface.[33]

If, in a region of fluid, there operates a mechanism by which an *external* force acts on the fluid, a new term must be added to the linearised momentum equations (3.12);

$$\partial p/\partial x - f_x = -\rho_0\, \partial u/\partial t \qquad (3.15a)$$

$$\partial p/\partial y - f_y = -\rho_0\, \partial v/\partial t \qquad (3.15b)$$

$$\partial p/\partial z - f_z = -\rho_0\, \partial w/\partial t \qquad (3.15c)$$

where f_i represents the component of an external force vector **f** per unit volume in direction i. In deriving the wave equation (3.13), the momentum equations are differentiated with respect to their corresponding spatial co-ordinate (e.g. eqn (3.12a) is differentiated with respect to x). Consequently the effective acoustic source term is $\partial f_x/\partial x + \partial f_y/\partial y + \partial f_z/\partial z$, which is the divergence of **f**.

The most common form of physical source in this second category is an unsteady flow of fluid over a rigid surface, such as a turbulent boundary layer, or the impingement of an air jet on a solid body: the edges of such bodies, especially if sharp, form particularly effective sources (try blowing on the edge of a piece of card). Clearly, the surfaces themselves cannot constitute active radiators; being rigid, they can do no work on the fluid, and hence can radiate no sound power. However, their presence serves to constrain the non-acoustic, hydrodynamic fluid motions, thereby producing momentum changes, and associated density changes, which would not occur in the absence of the solid body: a fraction of the kinetic energy of the flow is converted into sound energy. This second category of source is far less efficient at converting mechanical energy into sound than those in the first category. Certain sources which appear actively to displace fluid volume may actually be considered to fall into this category, rather than the first. In these, the *net* instantaneous volumetric displacement is zero at all times, but they produce fluctuations of local fluid momentum, which, of course, requires a force. Any oscillating body having dimensions small compared with an acoustic wavelength generates sound by this mechanism. A violin string itself generates a very small amount of sound in this manner which is in no way commensurate with the sound radiated by the body.

The third category of sources includes those associated with unsteady fluid flow in the absence of solid bodies, such as turbo-jet exhausts. These sources cannot be explained on the basis of linearised fluid dynamic equations and the linear approximation to the relationship between acoustic pressure and density fluctuations (eqn (3.5)). In view of the complexity of hydrodynamic noise source theory, the reader is directed to Refs 34 and 35 for authoritative expositions of the subject. This type of source is the least efficient of all— fortunately for the development of the civil airliner business!

3.7 SOLUTIONS OF THE WAVE EQUATION

As explained above, the wave equation takes various forms, depending upon the co-ordinate system used in its derivation. In any particular case, the choice is largely dictated by the geometric form of the boundaries of the fluid, and of the source distribution. In most practical cases there is no exact analytical solution of the equation, because sources and boundaries are geometrically so complicated. However, it is useful to study a number of simple examples, the solutions of which assist the understanding and approximate modelling of complicated physical systems. Considerable emphasis is placed on the particle velocity fields because sound energy flow is produced by 'co-operation' between sound pressure and particle velocity.

3.7.1 The plane wave

The sound field inside a long, uniform tube with rigid walls is constrained by the presence of the walls to take a particularly simple form at frequencies for which the acoustic wavelength exceeds about half the peripheral length of the tube cross-section (the exact ratio depends upon the cross-section geometry). At any instant of time, each acoustic variable is uniform over any plane perpendicular to the tube axis, irrespective of the time dependence of the field. Such a field is known as a 'plane wave field'. The governing homogeneous wave equation may be reduced to its one-dimensional form

$$\partial^2 p/\partial x^2 - (1/c^2)\partial^2 p/\partial t^2 = 0 \tag{3.16}$$

where $c^2 = \gamma P_0/\rho_0$. The general solution of this equation is

$$p(x, t) = f(ct - x) + g(ct + x) \tag{3.17}$$

where f and g are functions which depend on the spatial and temporal boundary conditions which obtain in any particular case. It is clear from the forms of the arguments of these functions that they represent disturbances travelling at speed c in the positive-x and negative-x directions, respectively.

The associated particle velocity distribution may be obtained by applying the momentum eqn (3.12a) to eqn (3.17), to give

$$u(x, t) = (1/\rho_0 c)f(ct - x) - (1/\rho_0 c)g(ct + x) \tag{3.18}$$

The quantity $\rho_0 c$ is termed the 'characteristic specific acoustic impedance' of the fluid. Equation (3.18) shows that the particle velocity can only be obtained from the pressure if *both* f and g are known. This is why attempts are made to devise anechoic terminations for ducts designed for the determination of in-duct sound power: one of the two waves is suppressed, and a single pressure microphone can be used to estimate particle velocity, and hence, sound intensity. The relationship between the pressure and particle velocity in the two individual waves is clearly independent of frequency.

Note that g is related to f in any bounded region solely by the relationship between pressure and particle velocity at *one* limiting boundary, irrespective of conditions at the other limit. This fact forms the basis of the measurement of sample acoustic impedance in a standing wave tube.

The plane *progressive* wave model (f or g alone) is an extremely useful idealisation which is hardly ever realised in practice outside the laboratory. However, just like the concept of the unrealisable eternal single frequency signal, which is so essential to complex signal analysis, the plane progressive wave may be utilised mathematically to analyse sound fields of any degree of spatial complexity, and to form the basis of spatial filtering techniques utilised in transducer arrays of various forms.

3.7.2 The spherical wave

The acoustic wave equation may be developed in a spherical co-ordinate system from first principles, or it may be obtained by transformation of the rectangular Cartesian form, eqn (3.13): the details will be found in suitable textbooks. Co-ordinates x, y and z are replaced by r, θ and ϕ, which are, respectively, the radial co-ordinate, the azimuthal angle and the angle of declination.

The equation in terms of pressure is

$$\partial^2 p/\partial r^2 + (2/r)\partial p/\partial r + (1/r^2 \sin \theta)\partial/\partial\theta(\sin \theta(\partial p/\partial\theta))$$
$$+ (1/r^2 \sin^2 \theta)\partial^2 p/\partial\phi^2 - (1/c^2)\partial^2 p/\partial t^2 = 0 \tag{3.19}$$

The associated particle accelerations are obtained by the application of the fluid momentum equations in the radial, and two orthogonal tangential directions:

$$\partial p/\partial r = -\rho_0 \partial u_r/\partial t \tag{3.20a}$$

$$(1/r)\partial p/\partial\theta = -\rho_0 \partial u_\theta/\partial t \tag{3.20b}$$

$$(1/r)\partial p/\partial\phi = -\rho_0 \partial u_\phi/\partial t \tag{3.20c}$$

A special case of fundamental importance is the spherically symmetric field, which may be used analytically to construct the fields generated by spatially extended complex sources. By suppressing the θ- and ϕ-dependent terms eqn (3.19) becomes

$$\partial^2(pr)/\partial r^2 - (1/c^2)\partial^2(pr)/\partial t^2 = 0 \qquad (3.21)$$

which is seen to be analogous to eqn (3.13), with a dependent variable (pr), the product of the pressure and the radius: it does not apply at the origin $r = 0$. The general solution of eqn (3.21) is

$$p(r, t) = (1/r)[f(ct - r) + g(ct + r)] \qquad (3.22)$$

where f and g are functions which depend upon the spatial and temporal boundary conditions. The second term in the square bracket represents a disturbance approaching the origin from an infinite distance, and is generally of little practical significance. The first term represents a disturbance which decreases in strength inversely with distance from the origin. The particle accelerations and velocities are purely radial in direction. At this stage it is appropriate to specialise to simple harmonic time dependence because the relationship between particle velocity and pressure in a spherically spreading wave field is frequency-dependent. The pressure field may be expressed in complex exponential form by

$$p(r, t) = (A/r)\exp[i(\omega t - kr)] \qquad (3.23)$$

in which A is (mathematically) complex, to accommodate arbitrary phase. The symbol k represents the acoustic wavenumber, which represents the rate of change of phase with distance, in the same way that circular frequency ω represents the rate of change of phase with time; wavenumber k can therefore be considered as 'spatial frequency'. Only the real part of such expressions is physically meaningful, but linear operations may be performed on the whole expression without resorting first to the extraction of the real part. Application of the momentum equation yields the relationship between pressure and radial partial velocity u_r,

$$u_r(r, t) = (A/\omega\rho_0 r)(k - i/r)\exp[i(\omega t - kr)] = [p(r, t)/\rho_0 c](1 - i/kr) \quad (3.24)$$

Although this field is a function of only one space variable, it is clearly more complicated than the plane wave field because the phase relationship between pressure and particle velocity depends upon distance, through the non-dimensional parameter kr. The field may be divided into two regions: (i) a near field in which $kr \ll 1$, and the particle velocity is nearly in quadrature with the pressure, and varies as r^{-2}; (ii) a far field in which $kr \gg 1$, the particle velocity is nearly in phase with the pressure, and varies as r^{-1}.

Mathematical analysis of relationships between harmonically varying quantities is most conveniently done in terms of complex amplitudes, of which the general form is $X\exp(i\omega t)$. For example, the complex amplitude of u_r is $(A/\omega\rho_0 r)(k - i/r)\exp(-ikr)$. The time-average of the product of two harmonically varying quantities represented by $X \exp(i\omega t)$ and $Y \exp(i\omega t)$ over one period $2\pi/\omega$ is given by $\frac{1}{2}\mathrm{Re}\{X Y^*\}$, where $*$ indicates 'complex conjugate'. For example, the mean square particle velocity is

$$\overline{u_r^2} = [\overline{p^2}/(\rho_0 c)^2](1 + (1/kr)^2)$$

In the near field, the ratio of mean square particle velocity to mean square pressure is seen to exceed the plane wave ratio by a factor $(kr)^{-2}$: this is a characteristic feature of near fields in general.

The specific acoustic impedance of a harmonic acoustic field is defined as the ratio of the complex amplitudes of pressure and particle velocity. In this case it is given by

$$z = (\rho_0 c)[(1 + i/kr)/(1 + (1/kr)^2)] \tag{3.25}$$

In mechanical terms, the imaginary, or reactive component of this impedance, being positive, has the characteristic of inertia. This immediately suggests that some kinetic energy is stored in the near field, and not propagated away to infinity: later we shall see that this is indeed the case. The reason for the difference between the relationship between pressure and particle velocity in the spherical field and in the plane wave field is the radial pressure gradient associated with spherical spreading.

As already indicated, the elementary spherical wave solution to the wave equation may be applied to the analysis of fields produced by complicated spatial source distributions. This is done by representing a source as an assemblage of vanishingly small elementary sources, and expressing the total field as a summation of the associated sound fields, due account being taken of time delay, or its equivalent in the frequency domain, phase. (Such superposition may, of course, only be validly applied to fields in which the linearisation approximations apply.) This procedure is expressed formally in terms of Green's functions, of which the most general form, the so-called 'free space' Green's function, corresponds to eqn (3.23) when the field is generated by a spatially concentrated 'point' source of fluctuating mass injection (source type (i) of Section 3.6.1).

Such a 'point' source may be represented by a sphere of radius a, pulsating uniformly at a frequency at which its non-dimensional radius $ka \ll 1$. If the surface of the sphere is supposed to have a normal velocity $U \exp(i\omega t)$, and the expression for radial particle velocity (eqn (3.24)) is equated to this surface velocity, the coefficient A in eqn (3.23) is given by

$$A = i\omega\rho_0 Q/4\pi \tag{3.26}$$

where the volumetric velocity of the source Q is defined to be the rate of outflow of fluid volume, of which the amplitude is $4\pi a^2 U$. Hence the pressure field may be expressed as

$$p(r, t) = i\omega\rho_0 Q G(kr) \exp(i\omega t)$$

where $G(kr) = (1/4\pi r)\exp(-ikr)$ is the free space Green's function. As already mentioned, the source strength is characterised by $\partial q/\partial t$; in this case we may identify the harmonic source strength as $i\omega Q$.

3.7.3 Interference fields

It is possible mathematically to synthesise any source distribution, of whatever physical nature, in the form of an array of point volumetric sources of appropriate strength and phase, although it is not necessarily the most suitable or useful form of representation in all cases. In terms of such a representation, any field except that of an isolated point source constitutes an interference field. The relationship between pressure and particle velocity normally varies greatly with position in an interference field, as the following simple example shows.

Consider two point volumetric sources (also termed 'point monopoles') of strengths Q and $-Q$, of the same frequency, some distance apart, as shown in Fig. 3.5. The complex amplitude of pressure is

$$P(r) = (i\omega\rho_0 Q/4\pi)[(1/R_1)\exp(-ikR_1) - (1/R_2)\exp(-ikR_2)] \quad (3.27)$$

The particle velocities associated with these two terms are directed along radii centred on the respective sources; they are given by eqn (3.24). The relative phase of the two components is equal to $\pi + k(R_1 - R_2)$. The instantaneous particle velocity is the vector sum of the instantaneous particle velocities produced by each source in isolation (principle of linear superposition) as shown by Fig. 3.5. Where these two components are neither in phase, nor in anti-phase, the particle velocity vector will rotate through 2π during one cycle, and the fluid particle trajectory will take the form of an ellipse (of which a Lissajous figure on an oscilloscope is an analogy): rotation will therefore occur everywhere except on the plane of symmetry. It is difficult to appreciate

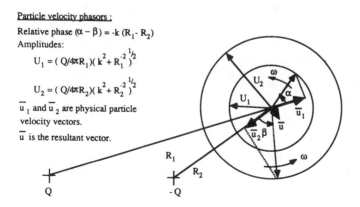

Fig. 3.5. Rotation of particle velocity phasors causes the resultant velocity vector to rotate (unless $\alpha - \beta = 0$ or π).

the physical significance of this conclusion in terms of the complicated form of the expression for pressure in eqn (3.27), and the even more complicated expression for the vector particle velocity. Therefore, let us consider a simpler model in which the two quantities are represented by arbitrary simple harmonic functions.

Figure 3.6 shows two vectors v_1 and v_2 which represent the instantaneous particle velocities produced by two sources lying in the same plane. These two velocities are expressed in complex exponential form as

$$v_1 = V_1 \exp[i(\omega t + \phi_1)]$$

and

$$v_2 = V_2 \exp[i(\omega t + \phi_2)]$$

and are represented by phasors in Fig. 3.6. The real parts of these expressions are

$$v_1 = V_1[\cos(\omega t)\cos(\phi_1) - \sin(\omega t)\sin(\phi_1)]$$

and

$$v_2 = V_2[\cos(\omega t)\cos(\phi_2) - \sin(\omega t)\sin(\phi_2)]$$

The total velocity vector is given by

$$\mathbf{v} = [v_1\cos(\theta_1) + v_2\cos(\theta_2)]\mathbf{i} + [v_1\sin(\theta_1) + v_2\sin(\theta_2)]\mathbf{j} \qquad (3.28)$$

This vector clearly rotates, unless $\phi_1 - \phi_2 = 0$ or π, since the ratio of the two vector components in eqn (3.28) is not independent of time. Velocity vector rotation will occur in any harmonic sound field generated by an *extended* source, unless the observation point is so far away from the source that it subtends a negligible angle at that point, in which case the particle velocity vectors produced by each element of the source have negligibly different

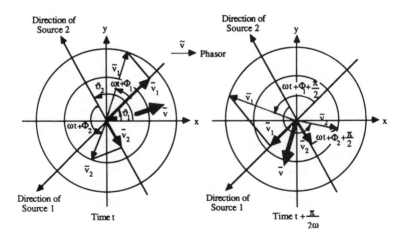

Fig. 3.6. Illustration of the rotation of particle velocity vector (no rotation if $\phi_1 - \phi_2 = 0$ or π).

directions. In a field generated by a source of aperiodic time dependence, the direction of the particle velocity vector at a point will exhibit irregular behaviour.

Consideration of the time-varying direction of fluid particle velocity raises the question of the physical interpretation of the specific acoustic impedance at any point in such an interference field. This quantity is defined as the ratio of the complex amplitude of any one frequency component of pressure at that point to the complex amplitude of the associated particle velocity component (see Section 3.8). It is easier to put a physical interpretation on this quantity if the direction of the associated velocity component is specified, as it is in the case of the normal surface impedance of a plane surface.

The reflection or scattering of sound waves by bodies or surfaces of impedance different from that of the wave-bearing fluid, produces interference between the incident and reflected, or scattered, field components. The phenomenon is most clearly observed in single frequency fields, where it is revealed by the presence of stationary wave patterns. It is of great practical significance in enclosures such as rooms and ducts, where it produces modal behaviour and characteristic frequencies associated with multiple reflection from the bounding surfaces. As one might expect from the previous discussion, the relationship between pressure and particle velocity in such fields is usually extremely complicated, as the following examples show.

The simplest example of an interference field is that produced by the superposition of two simple harmonic (SH) plane waves travelling in opposite directions, in which the pressure may be expressed as

$$p(x, t) = A \exp[i(\omega t - kx)] + B \exp[i(\omega t + kx)] \tag{3.29}$$

in which A and B are arbitrary complex amplitudes. The particle velocity is

$$u(x, t) = (A/\rho_0 c) \exp[i(\omega t - kx)] - (B/\rho_0 c) \exp[i(\omega t + kx)] \tag{3.30}$$

It is convenient to express the complex ratio B/A as $R \exp(i\theta)$. The ratio of pressure to particle velocity is defined to be the 'Specific Acoustic Impedance' z, of which the non-dimensional form is $z/\rho_0 c$.

$$z/\rho_0 c = [(1 - R^2 + 2iR \sin(2kx + \theta))/(1 + R^2 - 2R \cos(2kx + \theta))] \tag{3.31}$$

This expression shows that the interference fields produced by reflection of a normally incident plane wave by a plane rigid surface ($R = 1$), or by a plane pressure-release surface ($R = -1$ and $p = 0$), are wholly *reactive*, because the pressure is everywhere in quadrature with the particle velocity. (A water–air interface, seen from the water side, closely approximates a pressure-release surface.) If, as is always the case in practice, $|R| < 1$, the phase angle between pressure and particle velocity, and the magnitude of their ratio, varies with position x, as shown in Fig. 3.7.

Next, we consider an elementary example of a two-dimensional interference field in an infinitely long (or anechoically terminated), uniform, planar duct having rigid walls (Fig. 3.8). The physically possible forms of sound field in the

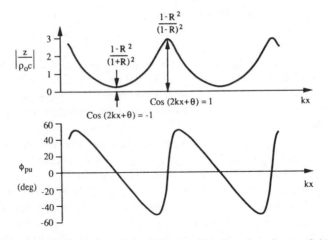

Fig. 3.7. $|z/\rho_0 c|$ and ϕ_{pu} as functions of distance in a plane interference field ($R = 0.5$, $\theta = 0.5$).

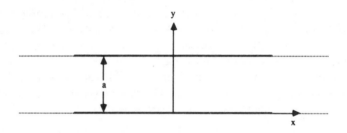

Fig. 3.8. A uniform, planar, rigid-walled duct.

duct are represented by solutions to the wave equation, subject to the boundary conditions of zero particle velocity normal to the walls; for simplicity, SH time dependence is again assumed. The SH form of the wave equation is, in two-dimensional rectangular Cartesian co-ordinates,

$$\partial^2 p/\partial x^2 + \partial^2 p/\partial y^2 + k^2 p = 0 \tag{3.32}$$

Let us assume that a wave can propagate along the duct with a wavenumber k_x, and seek a corresponding solution for the transverse distribution of the pressure, for which we assume a form $p(y) = P \exp(k_y y)$. The resulting algebraic equation for k_y is

$$k_y^2 = -(k^2 - k_x^2)$$

or

$$k_y = \pm i[(k^2 - k_x^2)^{1/2}] \tag{3.33}$$

Hence

$$p(x, y, t) = (A \exp(ik_y) + B \exp(-ik_y)) \exp[i(\omega t - k_x x)] \tag{3.34}$$

The rigid wall boundary conditions require $\partial p/\partial y$ to be zero for $y = 0$ and $y = a$. These conditions are simultaneously satisfied only if $A = B$ and $k_y = n\pi/a$. Equation (3.33) then demands that the axial wavenumber must satisfy the equation

$$k_x = k_n = \pm [k^2 - (n\pi/a)^2]^{1/2} \tag{3.35}$$

in which n is zero, or any integer. There is, therefore, an infinite number of solutions to the Helmholtz equation at any frequency, of which the general form is

$$p(x, y, t) = P_n \cos(n\pi y/a) \exp[i(\omega t - k_x x)] \tag{3.36}$$

where k_n is given by eqn (3.35).

Such a transverse distribution of pressure, which is characteristic of the geometry of the duct, and which is caused by multiple reflection of travelling waves from the duct walls, is known as a 'characteristic function', or duct 'mode': mode zero corresponds to axial plane wave propagation. A mode excited at any one frequency may only propagate along the duct with a unique wavenumber determined by eqn (3.35). For each mode there is a frequency at which k_x equals zero; this is termed the 'cut-off' frequency of that mode. The modes of propagation are characterised by their dispersion relationships between k_x and frequency, as illustrated by Fig. 3.9. A mode cannot propagate at frequencies below its cut-off frequency, when k_x is purely imaginary. Equations (3.35) and (3.36) show that the modal pressure will then decay exponentially with axial distance: the mode is said to be 'evanescent'.

The particle velocity components in the x- and y-directions associated with individual modes are respectively

$$u(x, y, t) = (1/\omega \rho_0) k_n P_n \cos(n\pi y/a) \exp[i(\omega t - k_x x)] \tag{3.37}$$

and

$$v(x, y, t) = -(i/\omega \rho_0)(n\pi/a) P_n \sin(n\pi y/a) \exp[i(\omega t - k_x x)] \tag{3.38}$$

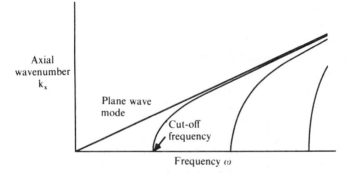

Fig. 3.9. Dispersion curves for planar duct modes.

The choice of sign in eqn (3.35), when $k < n\pi/a$, is dictated by the physical condition that the pressure in evanescent modes cannot increase with distance from the point of generation. Equation (3.37) shows that the axial particle velocity component is in phase with the local pressure in propagating modes, but is in quadrature with the pressure in evanescent modes. In contrast, the transverse particle velocity component is always in quadrature with the pressure.

A complete solution to the homogeneous Helmholtz equation consists of the sum of the infinite number of modal solutions. It will be realised that the relationship between pressure and particle velocity, and the time-dependent directional orientation of the velocity vector, can vary with frequency and with position in a duct in extremely complex manners.

The actual field generated by any particular acoustic source distribution in a duct depends upon the coupling of that source to the possible duct modes. For example, a two-dimensional line monopole source of volumetric velocity per unit length Q', located at $x_0 = 0$, y_0, will produce a modal pressure amplitude P_n given by

$$P_n = (\omega \rho_0 Q'/ak_n) \cos(n\pi y_0/a) \tag{3.39}$$

This expression suggests that the modal pressure becomes infinite at the cut-off frequency of that mode, when $k_n = 0$. In practice, the non-anechoic nature of duct terminations, or discontinuities, together with the finite internal impedance of real sources, mitigates this theoretically singular condition, although modal pressures do indeed reach high levels close to cut-off.

There is an alternative method of deriving an expression for the sound field generated by a point source operating within a volume enclosed by plane, rigid boundaries, without recourse to modal expansions. Reflections from the boundaries of the field propagating outwards from the source by the duct walls may be represented by an infinite set of image sources, as shown in Fig. 3.10.

One advantage of this representation is that the field close to the source is determined largely by the closer image sources, and the solution converges rapidly as the number of image sources is increased; by contrast, solutions based on modal synthesis converge extremely slowly. A further advantage is that the influence of modifications to the wall boundary conditions can be qualitatively understood more easily in terms of the image source construction. However, the field far along the duct axis from a source is better represented by modal synthesis because only the propagating modes need to be included; by contrast, the necessary number of image sources increases with axial distance from the source. The image analysis technique is widely used in theoretical studies of room acoustics.

Next we consider the acoustic field produced by the incidence of a plane harmonic wave on the mouth of a Helmholtz resonator, which consists of an enclosed volume of air that communicates with its surroundings via a channel, or neck. This is a good example of scattering of an incident field by the presence

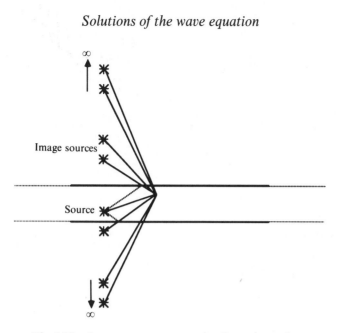

Fig. 3.10. Image source construction for a planar duct.

of a localised region of impedance which is very different from the characteristic impedance of the surrounding fluid. At resonance, this impedance approaches zero, resulting in very effective scattering and absorption of incident energy when the internal resistance of the resonator matches the radiation resistance experienced by the induced volumetric fluctuation at the mouth of the resonator. In this case the resonator can present an absorption cross-section hundreds of times the cross-section of the neck. How does it 'suck in' energy from such a large area of the incident wavefront? Figures 3.11(a–h) show the theoretical distribution of instantaneous particle velocity produced by the normal incidence of a plane wave upon a Helmholtz resonator set in an infinite rigid baffle. The distributions are frozen at instants $\frac{1}{8}$th of a period apart. (In order to accommodate the large range of magnitudes, the lengths of the plotted vectors are proportional to $u^{1/4}$.) Interference between the scattered and incident waves is seen to create a 'funnelling' effect. The associated intensity distributions presented in Chapter 4 reveal this phenomenon even more clearly.

Paradoxically, a resonator can only absorb strongly if it scatters strongly—hence the importance of matching of internal and radiation resistances for optimum absorber performance. Examples of this phenomenon of practical importance include scattering of underwater sound by marine animals, the influence of bubbles in liquids on sound propagation and the absorption of sound by flexible panels.

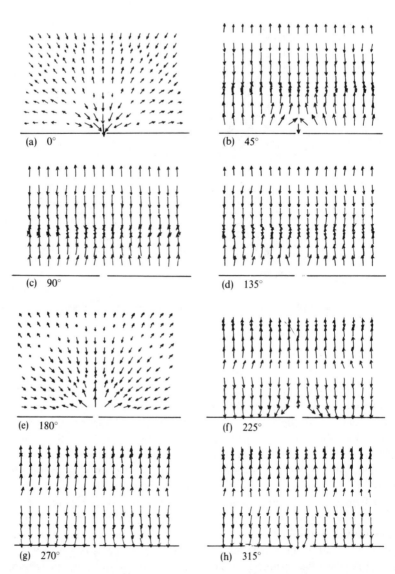

Fig. 3.11. Instantaneous particle velocity vectors at intervals of $\frac{1}{8}$ th period in the field of a plane wave normally incident on a Helmholtz resonator at the resonance frequency ($r_{\text{rad}}/r = 0.43$; vector scale $\propto u^{1/4}$).

3.8   SOUND RADIATION FROM VIBRATING SURFACES

A large proportion of sources of sound take the form of vibrating solid surfaces. The mechanism of production of sound is the reaction of a compressible medium to the normal component of acceleration of the surface with which it is in contact. Vibrating systems are very diverse in their geometric configurations, material properties and forms of construction; the sources of vibration are also extremely varied in their temporal and spatial characteristics. Idealisation, mathematical modelling and analysis of the audio-frequency vibrational behaviour of complicated assemblages of components are difficult and unreliable; still more difficult is the theoretical determination of the associated sound field which, in principle, requires knowledge of the amplitude and phase of the surface motion at all points, and at all frequencies of interest.

Irrespective of the details of a vibrating system, all sound fields so generated share certain physical features. Of considerable significance for the understanding and effective application of sound intensity measurement of such fields is the presence, in the close vicinity of a vibrating surface, of a near field. This is a region in which the fluid motion has a major component which closely approximates that which would exist if the fluid were incompressible: the associated field component satisfies the Laplace equation which takes the form of the wave equation when the speed of sound is put equal to infinity. The fluctuating velocities, and associated momentum densities, are far greater in proportion to the pressure than in the propagating far field, i.e. $|u| \gg |p/\rho_0 c|$. The presence in a near field of large pressure and particle velocity components which are not essentially associated with the transport of acoustic energy does not invalidate the principle of intensity measurement, but it puts an extra demand on the quality of the measurement system, which has to reject this 'non-propagating' component of the total field.

It has already been indicated that a spatially continuous vibrating surface may be represented acoustically by an array of elementary volumetric sources of appropriate amplitude and phase. Interference between the field components generated by these sources may create circulatory patterns of mean acoustic energy flow. These may exist as closed cells entirely within the fluid, or they may involve the vibrating structure as part of the circulation path. In the latter case, the surface exhibits regions of outward-flowing energy, interspersed with regions in which the energy flows into the surface. These regions are conventionally termed 'sources' and 'sinks'. This terminology is not considered to be entirely apt, since, unlike the incompressible fluid dynamic counterparts from which their names derive, the attribution does not essentially indicate *inexorable* outflow or inflow of energy. An acoustic 'source' can be turned into an acoustic 'sink' by the imposition of a sufficiently high local pressure in anti-phase with the local normal particle velocity; such pressure may, for example, be generated by vibrational motion elsewhere on the surface, and communicated to the 'source' region by acoustic wave motion. (This phenomenon may be demonstrated by setting up an array of three 'identical' small baffled loudspeakers in close proximity, and then measuring the intensity distribution as the polarity of the common pure tone signal fed to each unit is reversed in turn. The best effect is obtained at low frequencies where the array represents an acoustically compact source.)

This distinction between incompressible fluid 'sources' and 'sinks', and acoustic power 'sources' and 'sinks', highlights the fact that the latter involves the communication of disturbances over large distances, whereas, in the former case, local events are predominantly influenced by local conditions. The behaviour of the acoustic near field as an almost incompressible field suggests, correctly, that it is most strongly influenced by local conditions.

The nature of acoustic fields generated by vibrating surfaces will now be illustrated by an elementary idealised example. The model employed is a linear array of point monopoles having sources strengths and phases corresponding to a sinusoidal spatial distribution (Fig. 3.12): each source strength corresponds to the average over one-eighth of the array wavelength. The physical counterpart of this model is a simply supported flat strip, which is narrow compared with an acoustic wavelength, vibrating in a natural mode in a co-planar rigid baffle. Some examples of the instantaneous particle velocity distributions are illustrated in Figs 3.13(a–h), frozen at instants in a cycle of oscillation separated by intervals of $\frac{1}{8}$th period. (The plotted magnitude is proportional to $u^{1/4}$.) These clearly illustrate the rotation of the velocity vector in the field close to the source array. A little thought will reveal that the associated particle trajectories are ellipses.

As already mentioned, it is a rather complicated matter to predict theoretically the sound field generated by vibrating bodies of the degree of geometric complexity characteristic of such sources as musical instruments and industrial machinery. The main physical reason is that the field generated by an elementary volumetric source located on the surface of such an irregular body is not that produced by the source in free space, or when acting as part of a plane surface.[33] The presence of the rest of the (passive) surface of the body scatters the otherwise uniform, spherically spreading field of the individual source. Consequently, much effort has been devoted to the development of methods of numerical calculation of such sound fields, and to the associated computer algorithms (e.g. boundary element methods). In practice, the greater obstacle to prediction is the inadequacy of manageable idealisations of real structures for the purpose of estimating the distribution of surface normal vibration at the frequencies of greatest physiological sensitivity: indeed, no two

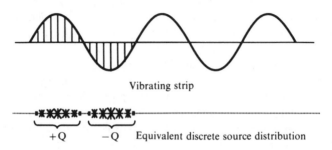

Fig. 3.12. Representation of a vibrating baffled strip by an array of discrete monopoles.

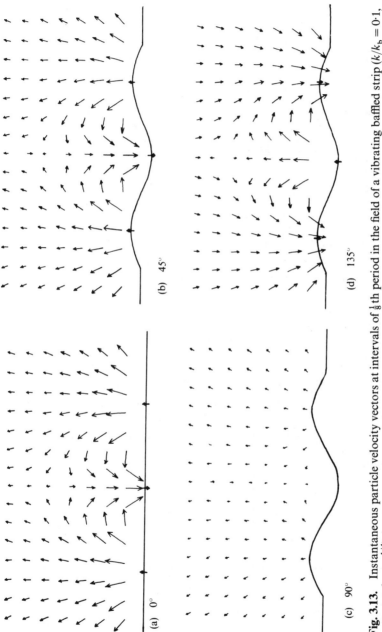

(a) 0°

(b) 45°

(c) 90°

(d) 135°

Fig. 3.13. Instantaneous particle velocity vectors at intervals of $\frac{1}{8}$th period in the field of a vibrating baffled strip ($k/k_b = 0.1$, vector scale $\propto u^{1/4}$).

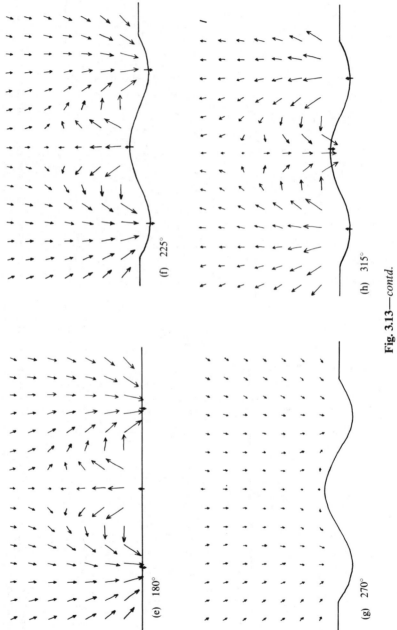

(e) 180°

(f) 225°

(g) 270°

(h) 315°

Fig. 3.13—*contd.*

nominally identical structures will exhibit closely matching amplitude and phase distributions at such frequencies. This is one of the reasons why the experimental investigation of radiation fields is such a useful tool for the noise control engineer. Unfortunately, the identification of an existing pattern of radiation does not necessarily indicate a practical solution, partly because of the interdependence of coherently vibrating surface elements in the process of sound generation (see Chapter 8 for further discussion).

3.9 ACOUSTIC IMPEDANCE

The specific acoustic impedance at a point in a sound field is defined as the ratio of the complex amplitude of an individual frequency component of sound pressure at that point, to the complex amplitude of the associated component of particle velocity. It is, of course, a complex number, giving the magnitude and phase of the ratio of the individual frequency components. It has an important bearing on energy flow in a sound field because the phase relationship between the pressure (a force-like quantity) and the fluid velocity clearly indicates the 'effectiveness of their co-operation', in the same way that the power factor does for voltage and current in an electrical circuit.

Since sound intensity probes necessarily indicate pressure and particle velocity component (implicitly in the case of the two-microphone probe), they may be used to indicate acoustic impedance anywhere in a sound field. This is of practical consequence for the determination of the sound absorption properties of material surfaces, which is most completely quantified by their surface normal specific acoustic impedance. Prior to the advent of instruments and probes for the measurement of sound intensity in the field, it was extremely difficult to evaluate the impedance of one surface in the presence of others; measurements were largely confined to the standing wave (impedance) tube in the laboratory. This has the disadvantage of restricting the samples to rather small areas, and to the plane wave frequency limit of the tube: it was impossible to investigate the properties of spatially non-uniform specimens, or to obtain reliable measurements of the impedance of surfaces *in situ*. Examples of results from *in situ* tests are presented in Chapter 10.

CHAPTER FOUR

Sound energy and sound intensity

4.1 WORK AND ENERGY

When a force acts upon a particle of matter, and that particle moves in space, work is done on the particle. The formal mathematical expression of this statement is

$$W = \int_A^B \mathbf{F}.\mathbf{ds} \tag{4.1}$$

in which $\mathbf{s}$ represents the particle path co-ordinate, and A and B represent the beginning and end points of the path. The instantaneous rate at which the force does work at any point on the path is

$$dW/dt = \mathbf{F}.(\mathbf{ds}/dt) = \mathbf{F}.\mathbf{u} \tag{4.2}$$

where $\mathbf{u}$ is the particle velocity vector. The scalar products of the vector quantities in eqns (4.1) and (4.2) indicate that work is done only by that component of the force in the direction of the particle displacement, or of the particle velocity, which is the same.

Substitution of Newton's Second Law of Motion into eqn (4.1) shows that the work done on a particle is equal to the change in a quantity defined to be the 'kinetic energy' of the particle; this is so, irrespective of the physical nature or origin of the force. The kinetic energy of a particle per unit mass is $\frac{1}{2}u^2$.

The forces which act on, and within, material media fall into one of two categories, depending upon whether or not any work which they do during displacements of the medium is recoverable by reversal of those displacements. They are defined, respectively, to be 'conservative' and 'non-conservative' forces. The most common examples of conservative force are those due to gravity, and the linear elastic forces produced by material strain. The non-conservative class includes the various forms of solid and fluid friction which convert mechanical work into heat. Negative work done by an internal conservative force is defined as 'potential energy'; this negative work is 'stored', and is recoverable by reversal of the work path. The sum of the kinetic and potential energies of a closed system on which only conservative internal forces act remains constant in the absence of external sources of energy input or extraction. Such systems are said to be 'conservative'.

4.2 SOUND ENERGY

In the inviscid gas model assumed in elementary acoustic theory, the only internal force is the pressure which arises from volumetric strain. Pressure is a manifestation of the rate of change of momentum of the gas molecules produced by their mutual interactions during random motion. Gas temperature is a manifestation of the kinetic energy density of translational molecular motion. The relationship between changes of pressure and density during volumetric strain depends upon the degree of heat flow between fluid regions of different temperature, or into, or out of, other media with which the gas is in contact. During small audio-frequency disturbance of real gases, heat flow is negligible in the body of the gas remote from solid boundaries, and the influence of irreversible changes due to fluid viscosity and molecular vibration phenomena can, to a first approximation, be neglected. The corresponding adiabatic bulk modulus represents a conservative elastic process.

The kinetic energy of a fluid per unit volume, symbolised by T, is clearly equal to $\frac{1}{2}\rho u^2$, where u is the speed of the fluid particle motion. The potential energy associated with volumetric strain of an elemental fluid volume is equal to the negative work done by the internal fluid pressure acting on the surface of the elemental volume during strain. Since the total volume change is given by the integral over the surface of the normal displacement of the surface, the potential energy per unit volume is given by

$$dU = -P(dV/V) \qquad (4.3)$$

The total pressure P is the sum of the equilibrium pressure P_0 and the acoustic pressure p. Lighthill[35] shows that the action of P_0 is associated with the convection of acoustic energy by the fluid velocity, a contribution to energy transport which is very small and which is balanced out by another small term, and may therefore be neglected. Hence, eqn (4.3) may be reduced to

$$dU = -p(dV/V) \qquad (4.4)$$

Equations (3.3) and (3.5) give dV/V as $-d\rho/\rho$, and $p/(\rho - \rho_0)$ as c^2. Hence, $d\rho = dp/c^2$, and with the small disturbance assumption that $(\rho - \rho_0)/\rho_0 \ll 1$, eqn (4.4) becomes

$$dU = p\,dp/\rho_0 c^2 \qquad (4.5)$$

The potential energy per unit volume is hence

$$U = p^2/2\rho_0 c^2 = p^2/2\gamma P_0 \qquad (4.6)$$

The total mechanical energy per unit volume associated with an acoustic disturbance, known as the 'sound energy density', is

$$e = T + U = \tfrac{1}{2}\rho_0 u^2 + p^2/2\rho_0 c^2 \tag{4.7}$$

This expression is totally general and applies to any sound field in which the small disturbance criteria, and the zero mean flow condition, are satisfied.

4.3 PROPAGATION OF SOUND ENERGY: SOUND INTENSITY

In the introduction, a simple physical argument was advanced for the phenomenon of transport of sound energy. The vibrational potential and kinetic energies of fluid elements in the path of a transient sound wave are zero before the wave reaches them, and zero again after the wave has passed. Provided that no local transformation of energy into non-acoustic form has occurred, the energy which they temporarily possess while involved in the disturbance has clearly travelled onwards with the wave. We proceed to derive an expression for the energy balance of a small region of fluid in a general sound field, making the assumption that small dissipative forces may, to a first approximation, be neglected. We must be careful to exclude any elements which are acted upon by *external* forces which may do work on them; for example, elements in contact with vibrating surfaces. We also assume that *no sources or sinks of heat or work are present*, and that heat conduction is negligible. These latter assumptions imply that changes of internal energy of a fluid element, and the associated temperature changes, are produced solely by work done on the element by the surrounding fluid during volumetric strain. Since the internal forces are then conservative, the rate of change of the mechanical energy of a region of fluid must equal the difference between the rate of flow of mechanical energy in and out of the region.

On the basis of the definition of mechanical work presented in Section 4.1, the rate at which work is done on fluid on one side of any imaginary surface embedded in the fluid, by the fluid on the other side, is given by the scalar product of the force vector acting on that surface times the normal fluid particle velocity vector through the surface. The rate of work is therefore expressed mathematically as

$$dW/dt = \mathbf{F}.\mathbf{u} = p\,\delta\mathbf{S}.\mathbf{u} \tag{4.8}$$

where $\delta\mathbf{S}$ is the elemental vector area which can be written as $\delta S\mathbf{n}$, where $\mathbf{n}$ is the unit vector normal to the surface, directed into the fluid receiving the work (Fig. 4.1). The work rate per unit area may be written

$$(dW/dt)/\delta S = pu_n \tag{4.9}$$

where $u_n = \mathbf{u}.\mathbf{n}$ is the component of particle velocity normal to the surface.

We define the *vector* quantity $p\mathbf{u}$ to be the instantaneous *sound intensity*, symbolised by $\mathbf{I}(t)$, of which the component normal to any chosen surface

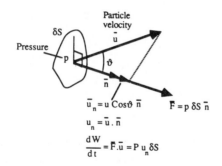

Fig. 4.1. Force on, and velocity through, a surface element in a fluid.

having unit normal vector **n** is $I_n(t) = \mathbf{I}(t).\mathbf{n}$. Note that, in general, both the magnitude and direction of $\mathbf{I}(t)$ at any point in space vary with time.

We may now express the energy balance of a region of fluid volume in terms of the flow of sound energy into and out of it. For simplicity, consider first a region in a two-dimensional sound field, shown in Fig. 4.2, in which the particle velocity vector has components u and v in the x- and y-directions respectively.

The rate of inflow of energy per unit depth is $[pu][\delta y] + [pv][\delta x]$: the rate of outflow is

$$[p + (\partial p/\partial x)\delta x][u + (\partial u/\partial x)\delta x][\delta y] + [p + (\partial p/\partial y)\delta y][v + (\partial v/\partial y)\delta y][\delta x]$$

Whereas it was appropriate to retain only first-order terms in the derivation of the wave equation, such a procedure is not appropriate to energy equations because all the terms would then disappear: hence we retain second-order terms, but neglect terms of higher order. The remaining expression for the net rate of outflow of energy per unit depth is $\partial/\partial x(pu) + \partial/\partial y(pv)$.

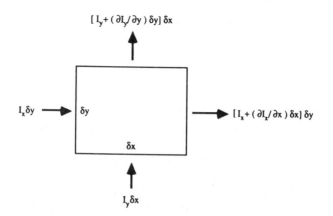

Fig. 4.2. Sound energy flux through the boundaries of a planar control volume in a fluid.

The rate of change of energy density of the fluid region is, from eqn (4.7), and using eqns (3.5), (3.11) and (3.12),

$$\partial e/\partial t = - [\partial/\partial x(pu) + \partial/\partial y(pv)] \tag{4.10}$$

The expected energy balance is therefore confirmed. A simple extension to a three-dimensional rectangular region of fluid gives the general relationship

$$\nabla.\mathbf{I}(t) = - \partial e/\partial t \tag{4.11a}$$

in which ∇ is the vector operator which, through the scalar product, generates the 'divergence' of the operand. In rectangular Cartesian co-ordinates this vector operator is expressed explicitly as

$$(\partial/\partial x)\mathbf{i} + (\partial/\partial y)\mathbf{j} + (\partial/\partial z)\mathbf{k}$$

so that when it operates on $\mathbf{I} = I_x\mathbf{i} + I_y\mathbf{j} + I_z\mathbf{k}$, the result is

$$\partial I_x/\partial x + \partial I_y/\partial y + \partial I_z/\partial z$$

If a sound field is time-stationary, the time-integral of eqn (4.11a) will converge to zero as the integration time extends beyond the period of lowest frequency component present. If work is being done on the fluid region by some external agent at a rate of W' per unit volume then eqn (4.11a) becomes

$$\nabla.\mathbf{I}(t) = - \partial e/\partial t + W' \tag{4.11b}$$

4.4   SOUND INTENSITY IN PLANE WAVE FIELDS

The relationship between instantaneous pressure and instantaneous particle velocity in a one-dimensional plane wave interference field is given by eqns (3.17) and (3.18) as

$$u^+ = p^+/(\rho_0 c) \quad \text{and} \quad u^- = - p^-/(\rho_0 c)$$

where the superscripts refer to the components propagating in the positive and negative x-directions. Hence, eqn (4.9) gives the instantaneous sound intensity as

$$I(t) = [(p^+)^2 - (p^-)^2]/(\rho_0 c) \tag{4.12}$$

in which the x- and t-dependence of the pressures is implicit. The time-averaged, or mean value of I in a time-stationary field is given by eqn (4.12) with the squares of instantaneous pressures replaced by mean square pressures.

Even in this most elementary of sound fields it is clearly not possible to measure sound intensity with a pressure microphone at one fixed position, because it cannot distinguish between the pressures associated with the two wave components travelling in opposite directions. It is clear from eqn (4.12) that, if the mean intensity is zero anywhere, it is zero at all positions, because both mean square pressures are independent of position. However, even if the mean is zero, the instantaneous intensity at any point fluctuates about this mean, indicating that energy is flowing to and fro in each local region. At certain times and places, the component wave pressures will be of the same sign and rather similar in magnitude, and the particle velocity will be correspondingly small; alternatively, the pressures can be similar in magnitude, but opposite in sign, and the total pressure will be small, while the particle velocity will be large. The conclusion must be that in any local region there is a continuous interchange between potential and kinetic energy, on which there may be superimposed a mean flow of energy through the region. This phenomenon may be understood more clearly by consideration of the simple harmonic plane wave interference field.

Consider first the pure *progressive* plane wave represented by $p(x, t) = A \cos(\omega t - kx + \phi)$. Equation (4.7) shows that the kinetic and potential energy densities are equal to each other at all times and positions;

$$e_k(x, t) = e_p(x, t) = [A^2/2\rho_0 c^2]\cos^2(\omega t - kx + \phi) = e/2$$

The instantaneous intensity is given by eqn (4.12) as

$$I(x, t) = [A^2/\rho_0 c]\cos^2(\omega t - kx + \phi) = ce \qquad (4.13)$$

Hence, $I/e = c$ for all x and t. The mean intensity $\bar{I} = \frac{1}{2}[A^2/\rho_0 c] = c\bar{e}$. The spatial distributions of instantaneous energies and intensity are illustrated in Fig. 4.3. It is seen that the energy is concentrated in 'clumps', spaced periodically at half-wavelength intervals. As indicated by eqn (4.13), the intensity at any point varies with time, but at no time takes a negative value; consequently it cannot be said to oscillate.

Now consider a pure *standing wave* in which the pressure takes the form $p(x, t) = 2A \cos(\omega t + \phi_1)\cos(kx + \phi_2)$. The spatial distribution of the kinetic

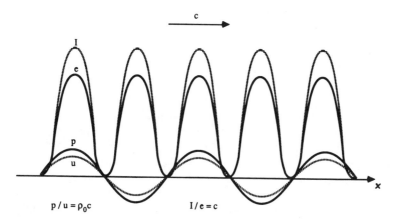

Fig. 4.3. Instantaneous spatial distributions of sound pressure, particle velocity, energy density and intensity in a plane progressive wave.

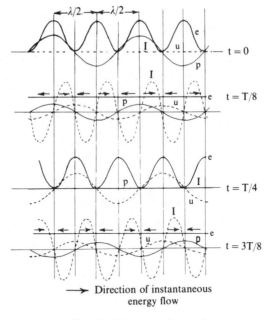

— Direction of instantaneous
energy flow

Fig. 4.4. Instantaneous spatial distributions of sound pressure, particle velocity, energy density and intensity in a pure standing wave at intervals of $\frac{1}{8}$th period.

and potential energy densities are shown at time increments of $\frac{1}{8}$th of a period in Fig. 4.4.

The distribution of instantaneous total energy density is given by

$$e(x,t) = [A^2/\rho_0 c^2][1 + \cos(2\omega t + 2\phi_1)\cos(2kx + 2\phi_2)] \qquad (4.14)$$

and the instantaneous intensity distribution is given by

$$I(x,t) = [A^2/\rho_0 c][\sin(2\omega t + 2\phi_1)\sin(2kx + 2\phi_2)] \qquad (4.15)$$

In this case $I/e \neq c$ and $\bar{I} = 0$, as indicated by Fig. 4.4. The instantaneous intensity expressed by eqn (4.15) represents a purely *oscillatory* flow of sound energy between alternating concentrations of kinetic and potential energy.

In all time-stationary acoustic fields, except the very special case of the plane travelling (progressive) wave, the instantaneous intensity may be split into two components: (i) an *active* component, of which the time-average (mean) value is non-zero, corresponding to local net transport of sound energy; and (ii) a *reactive* component, of which the time-average value is zero, corresponding to local oscillatory transport of energy. At any one frequency, these two intensity components are associated with the components of particle velocity which are, respectively, in phase and in quadrature with the acoustic pressure. The presence of local active intensity in a field does not necessarily imply that there is net transport of energy throughout an extended region of that field. As we shall see later, two- and three-dimensional interference fields contain local regions in which the active component of the intensity vector takes both positive and negative signs (opposite directions).

In all physical sound fields there takes place some dissipation of mechanical energy into heat; there must therefore exist a net flow of energy into those regions in which the dissipative mechanisms act, which, in steady sound fields, must be balanced by a corresponding net flow of energy out of those source regions in which it is generated. Consequently, purely reactive fields such as pure standing waves cannot exist; neither can the ideal diffuse field, in which plane waves of random phase and uniform amplitude are supposed to pass through every field point with equal probability from all directions, thereby producing zero net energy transport.

A general means of identifying the active and reactive components in a one-dimensional, *single frequency* sound field may be derived by considering the pressure and particle velocity in an arbitrary, plane interference field. Let us represent the pressure by $p(x, t) = P(x) \exp[i(\omega t + \phi_p(x))]$, in which $P(x)$ is the (real) space-dependent amplitude and $\phi_p(x)$ is the space-dependent phase. Hereinafter the explicit indication of the dependence of P and ϕ on x will be dropped for typographical clarity. The pressure gradient is

$$\partial p/\partial x = [dP/dx + i(d\phi_p/dx)P] \exp[i(\omega t + \phi_p)].$$

The momentum equation (3.12a) gives the particle velocity as

$$u = (i/\omega\rho_0)\partial p/\partial x = (1/\omega\rho_0)[-P(d\phi_p/dx) + i(dP/dx)] \exp[i(\omega t + \phi_p)].$$

The component of particle velocity in phase with the pressure is associated with the active component of intensity, which is given by their product as

$$I_a(x, t) = -(1/\omega\rho_0)[P^2(d\phi_p/dx)]\cos^2(\omega t + \phi_p) \qquad (4.16)$$

of which the mean value is

$$\bar{I}_a(x) = -(1/2\omega\rho_0)[P^2(d\phi_p/dx)] \qquad (4.17)$$

The component of particle velocity in quadrature with pressure is associated with the reactive component of intensity, which is given by their product as

$$I_{re}(x, t) = -(1/4\omega\rho_0)[dP^2/dx]\sin 2(\omega t + \phi_p) \qquad (4.18)$$

of which the mean value is zero.

We see that the active component of intensity is proportional to the spatial gradient of phase, and the reactive component is proportional to the spatial gradient of mean square pressure. Wavefronts, which are surfaces of uniform phase, lie perpendicular to the direction of the active intensity vector.

We may gain further insight into the nature of one-dimensional intensity fields by considering an example previously encountered in Section 3.7.3. The sound absorption properties of small samples of material are commonly measured in a 'standing wave' or 'impedance' tube below the lowest cut-off frequency of the tube. Suppose that the sample has a complex pressure reflection coefficient represented by $R \exp(i\theta)$. The pressure field is represented in complex exponential form by

$$p(x, t) = A\{\exp[i(\omega t - kx)] + R \exp(i\theta) \exp[i(\omega t + kx)]\}$$

which may be expressed in the general form introduced above

$$p(x, t) = P \exp(i\phi_p)\exp(i\omega t)$$

where

$$\phi_p = \tan^{-1}[(R \sin(kx + \theta) - \sin kx)/(R \cos(kx + \theta) + \cos kx)] \quad (4.19)$$

and

$$P^2 = A^2[1 + R^2 + 2R \cos(2kx + \theta)] \quad (4.20)$$

The spatial gradients of these quantities are

$$d\phi_p/dx = k[R^2 - 1]/[1 + R^2 + 2R \cos(2kx + \theta)]$$
$$= k(R^2 - 1)A^2/P^2 \quad (4.21)$$

and

$$dP^2/dx = -4A^2 kR \sin(2kx + \theta) \quad (4.22)$$

Observations in impedance tubes confirm that the spatial gradient of phase is greatest at pressure minima, and smallest at pressure maxima, as eqn (4.21) indicates, and can exceed that in a plane progressive wave.

Substitution of the above expressions into eqns (4.16)–(4.18) yields the following expressions for time-dependent active intensity $I_a(t)$, mean active intensity $\bar{I}_a$ and reactive intensity $I_{re}(t)$, respectively:

$$I_a(x, t) = (A^2/\rho_0 c)(1 - R^2)\cos^2(\omega t + \phi_p) \quad (4.23)$$
$$\bar{I}_a = (A^2/2\rho_0 c)(1 - R^2) \quad (4.24)$$

and

$$I_{re}(x, t) = (A^2/\rho_0 c)R \sin(2kx + \theta)\sin 2(\omega t + \phi_p) \quad (4.25)$$

The total instantaneous intensity is the sum of $I_a(t)$ and $I_{re}(t)$:

$$I(x, t) = (A^2/\rho_0 c)[(1 - R^2)\cos^2(\omega t + \phi_p)$$
$$+ R \sin(2kx + \theta)\sin 2(\omega t + \phi_p)] \quad (4.26)$$

It is clear that the mean active intensity is independent of x and uniform along the length of the tube, as it must be in the absence of dissipation in the fluid, and the mean value of the reactive intensity is zero. The ratio of the magnitudes of reactive to active intensity varies with position; it has a maximum value of $R/(1 - R^2)$ at the positions of maximum and minimum mean square particle velocity, and a minimum value of zero at maxima and minima of mean square pressure, respectively.

It may, at first, be difficult to reconcile eqn (4.13) with (4.26) for a pure progressive wave ($R = 0$), although eqns (4.15) and (4.26) for a pure standing wave ($R = 1$) are clearly of the same form. The reason is that in eqn (4.26) the spatial phase dependence is implicit in $\phi_p(x)$. Reference to eqn (4.19) with $R = 0$ indicates that, in this case, $\phi_p(x) = -kx$, which corresponds to the dependence seen in eqn (4.13). The constant phase term in the latter is totally arbitrary, and if employed would naturally appear as a constant addition to ϕ_p.

4.5 COMPLEX INTENSITY: ONE-DIMENSIONAL HARMONIC SOUND FIELDS

The total instantaneous intensity given by the sum of expressions in eqns (4.16) and (4.18) may be written in the form

$$I(x, t) = \bar{I}_a(x)[1 + \cos 2(\omega t + \phi_p)] + I_{re}(x) \sin 2(\omega t + \phi_p) \qquad (4.27)$$

in which

$$\bar{I}_a(x) = -(1/2\omega\rho_0)P^2[d\phi_p/dx] \qquad (4.28)$$

and

$$I_{re}(x) = -(1/4\omega\rho_0)dP^2/dx \qquad (4.29)$$

where

$$p(x, t) = P(x)\exp(i\phi_p(x))\exp(i\omega t)$$

A mathematically more compact form of eqn (4.27), which is analogous to the complex exponential representation of harmonically varying quantities, is

$$I(x, t) = \text{Re}\{C(x)[1 + \exp(-2i(\omega t + \phi_p))]\} \qquad (4.30)$$

in which

$$C(x) = \bar{I}_a(x) + iI_{re}(x)$$
$$= I(x) + iJ(x) \qquad (4.31)$$

where C is known as the 'complex intensity'. The real part of C is the mean (active) intensity, and the imaginary part J may be considered as the *amplitude* of the reactive intensity. Henceforth in this book, $\bar{I}_a$ will usually be termed 'the mean intensity' and will be symbolised by I, since it represents the quantity which is most widely measured, and corresponds to the most commonly stated definition of sound intensity as the long-time-average rate of flow of sound energy per unit area of fluid.

The form of the expression in eqn (4.30) may be unfamiliar to readers, but may be seen to be similar to the phasor representation of the square of the pressure in a harmonic sound field:

$$p^2 = [\text{Re}\{P\exp(i\phi_p)\exp(i\omega t)\}]^2$$
$$= P^2 \cos^2(\omega t + \phi_p)$$
$$= (P^2/2)[1 + \cos 2(\omega t + \phi_p)]$$
$$= (P^2/2)\text{Re}\{1 + \exp[2i(\omega t + \phi_p)]\}$$

The complex nature of C is a reflection of the fact that the two agents of energy flow, pressure and particle velocity, are not necessarily in phase (whereas in the example above p is, of course, in phase with itself).

Consider the explicit form of instantaneous intensity as the product of pressure, $p(x, t) = P \exp(i\phi_p) \exp(i\omega t)$ and particle velocity, $u(x, t) = U \exp(i\phi_u) \exp(i\omega t)$, where ϕ_u is analogous to ϕ_p:

$$I(x, t) = PU \cos(\omega t + \phi_p) \cos(\omega t + \phi_u)$$

which may be written

$$I(x, t) = \tfrac{1}{2} PU[\cos(2\omega t + 2\phi_p + (\phi_u - \phi_p)) + \cos(\phi_p - \phi_u)]$$

$$= \tfrac{1}{2} PU[(\cos(2(\omega t + \phi_p)) \cos \phi_r + \sin(2(\omega t + \phi_p)) \sin \phi_r + \cos \phi_r]$$

$$= \text{Re}\{\tfrac{1}{2} PU \exp(i\phi_r)[1 + \exp(-2i(\omega t + \phi_p))]\} \tag{4.32}$$

in which $\phi_p - \phi_u$ has been replaced by ϕ_r. By analogy with eqn (4.30),

$$C = I + iJ = \tfrac{1}{2} PU \exp(i\phi_r) \tag{4.33a}$$

$$|C| = \tfrac{1}{2} PU \tag{4.33b}$$

$$I = \tfrac{1}{2} PU \cos \phi_r \tag{4.33c}$$

and

$$J = \tfrac{1}{2} PU \sin \phi_r \tag{4.33d}$$

If a complex amplitude representation is employed, i.e. $p(x, t) = P \exp(i\omega t)$ and $u(x, t) = U \exp(i\omega t)$, then

$$C = \tfrac{1}{2} PU^* \tag{4.34a}$$

$$|C| = \tfrac{1}{2} |P||U| \tag{4.34b}$$

$$I = \tfrac{1}{2} \text{Re}\{PU^*\} \tag{4.34c}$$

and

$$J = \tfrac{1}{2} \text{Im}\{PU^*\} \tag{4.34d}$$

The expressions for active and reactive intensity in an impedance tube (eqns (4.23) and (4.25)) may be rapidly obtained by the application of the above relationships, together with eqn (4.30).

Graphical representation of $I(x, t)$ according to either eqns (4.30) or (4.32), together with the time histories of p and u, are illustrated in Fig. 4.5. The rotational speed of the moving component of the total phasor is 2ω, and the instantaneous intensity is represented by the real component of the total phasor. This phasor representation may be extended to two- or three-dimensional sound fields simply by separate phasor representation for each orthogonal component of vector $\mathbf{I}(r, t)$. (It must be remembered when manipulating analytical expressions for pressure that the pressure phase, although arbitrary at any one position (because of arbitrary time origin), is a function of position, and therefore the difference between its value at different positions in space is not arbitrary.)

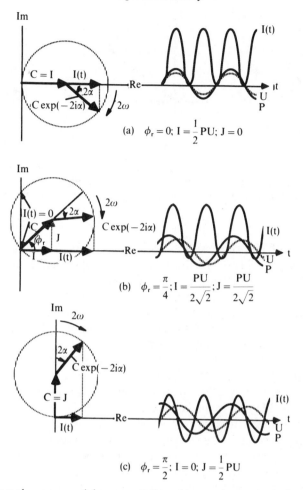

(a) $\phi_r = 0$; $I = \dfrac{1}{2}PU$; $J = 0$

(b) $\phi_r = \dfrac{\pi}{4}$; $I = \dfrac{PU}{2\sqrt{2}}$; $J = \dfrac{PU}{2\sqrt{2}}$

(c) $\phi_r = \dfrac{\pi}{2}$; $I = 0$; $J = \dfrac{1}{2}PU$

Fig. 4.5. Complex exponential representation of instantaneous intensity (the phasor rotates clockwise at speed 2ω).

It is seen from eqns (4.28), (4.29), (4.33) and (4.34) that the spatial distribution of complex intensity may either be derived from the spatial distributions of the mean square value and phase of the pressure, or from that of the root mean square values of pressure and particle velocity, together with their relative phase. The former derivation is more amenable to experimental determination, requiring as it does the transduction of only one type of physical quantity.

Anticipating Chapter 5 on Principles of measurement of sound intensity, in which the two-microphone method is explained, it is worth noting here that phasor representation of the two pressures measured at points separated by a small distance d in a harmonic field elucidates the physical interpretation of the mathematical relationships presented above (Fig. 4.6). The particle acceleration is in phase with the pressure difference, and the particle velocity is in

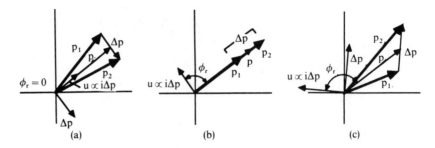

Fig. 4.6. Phasor representation of harmonic pressures at two closely spaced field points (all phasors rotate anti-clockwise at speed ω). (a) Plane progressive wave (u in phase with p). (b) Pure standing wave (u in quadrature with p). (c) General field (u leads p by ϕ_r).

quadrature with the acceleration: $\partial u/\partial t \approx -(1/\rho_0)(\delta p/d)$ and $u = (i/\omega\rho_0)\partial p/\partial x \approx (i/\omega\rho_0)(\delta p/d)$. Therefore the phase angle between the mean pressure $(p_1 + p_2)/2$ and the particle velocity approaches $\pi/2$ if $|p_1| - |p_2|$ is large, in which case the spatial gradient of mean square pressure is also large, indicating that the reactive intensity component is strong relative to the active component. When the magnitudes of the pressures are similar, the phase difference between the two pressures may be small, in which case both active and reactive components are small; such is the case at a pressure minimum in an impedance tube terminated by a strongly reflective sample.

In the majority of cases of practical interest to the engineer, sound fields contain many different frequency components. Although the Fourier spectrum of *instantaneous* intensity then contains sum and difference frequencies, the contributions of individual frequency components of pressure and particle velocity to the mean intensity are independent, because of their orthogonal properties; therefore they may simply be summed to give the mean intensity. Fortunately, one is not normally interested in the frequency spectrum of the instantaneous intensity, or there would be problems with the excessive averaging time required for the evaluation at small difference frequencies. We can, however, infer that local time-dependent energy flows of such fields have extremely long time scales, and therefore are very slow to settle down after the initiation of a field.

4.6 SOUND INTENSITY IN TWO- AND THREE-DIMENSIONAL HARMONIC SOUND FIELDS

4.6.1 Complex vector intensity

The foregoing analyses may be extended to any harmonic, three-dimensional sound field by replacing the position co-ordinate x by a position vector $\mathbf{r}$. The three orthogonally directed components of active and reactive intensity are then obtained by using the appropriate components of the spatial gradients of mean square pressure and pressure phase. It is, of course, appropriate to use vector notation to represent the quantity $\mathbf{I}(t)$. In a harmonic sound field the real and imaginary components of the complex intensity have vector properties, but their directions are not necessarily the same; however the time dependences remain the same as those expressed by eqns (4.16) and (4.18). Equation (4.33a) is extended to a vector form as

$$\mathbf{C} = \tfrac{1}{2}P[U\exp(i(\phi_p - \phi_u))\mathbf{i} + V\exp(i(\phi_p - \phi_v))\mathbf{j} + W\exp(i(\phi_p - \phi_w))\mathbf{k}] \quad (4.35)$$

where U, V and W are the amplitudes of the three particle velocity components in directions of the unit vectors $\mathbf{i}, \mathbf{j}, \mathbf{k}$. Equation (4.35) indicates that $\mathbf{I}$ and $\mathbf{J}$ are only similarly directed when all three ϕ_r have the value $\pi/4$.

The three-dimensional equivalents of eqns (4.31), (4.28) and (4.29) are

$$\mathbf{C}(\mathbf{r}) = \mathbf{I}(\mathbf{r}) + i\mathbf{J}(\mathbf{r}) \quad (4.36a)$$

$$\mathbf{I}(\mathbf{r}) = (1/2\omega\rho_0)P^2(\mathbf{r})\nabla\phi_p(\mathbf{r}) \quad (4.36b)$$

$$\mathbf{J}(\mathbf{r}) = -(1/4\omega\rho_0)\nabla P^2(\mathbf{r}) \quad (4.36c)$$

One interesting general conclusion is that spatial variations of the mean square pressure of individual frequency components in a multi-frequency field always indicate the presence of oscillatory energy flow even when the overall mean square pressure is spatially quite uniform, as in a broad band field in a large reverberation chamber. Another important implication of the proportionality between $\mathbf{J}$ and the spatial gradient of mean square pressure is that the fields of strongly directional sound sources must be strongly reactive at the 'edges' of the directivity lobes, especially fairly close to the source (see Ref. 38, p. 164).

The direction of the reactive component of intensity, which is opposite to that of the gradient of mean square pressure, is of practical significance; convergence of the reactive component vectors onto a point indicates a local region of low, or zero, acoustic pressure; divergence indicates a region of pressure concentration such as the near field of a point source (see Section 8.2).

The following examples[36] illustrate the complexity of two-dimensional intensity fields, and serve to introduce the phenomenon of apparent circulation of active intensity which occurs in many interference fields. Consider first two orthogonally directed plane waves represented by

$$p_1(x, y, t) = A\exp[i(\omega t - kx)]$$

and

$$p_2(x, y, t) = B\exp[i(\omega t - ky)]$$

The particle velocities in the x- and y-directions are, respectively,

$$u = p_1/\rho_0 c \quad \text{and} \quad v = p_2/\rho_0 c$$

Hence

$$C_x = (1/2\rho_0 c)[A^2 + AB\exp[ik(x - y)]] \quad (4.37a)$$

and

$$C_y = (1/2\rho_0 c)[B^2 + AB\exp[ik(y-x)]] \qquad (4.37b)$$

the corresponding expressions for I and J being given by the real and imaginary parts respectively. The total mean intensity vector, which is given by the vector sum of I_x and I_y, is represented in Fig. 4.7, in which $B = A/(2)^{1/2}$.

The complexity of spatial distribution of mean intensity created by the superposition of elementary wave fields is due to the co-operation between the pressure in each component field and the particle velocities in all the others. This results in a spatial 'modulation' of the sum of the active intensities of the individual fields. Integration of I_x over an integer number of intervals of y of $2\pi/k$ yields the sound power flux of wave 1 alone; similarly, integration of I_y over x yields the power flux of wave 2 alone. The local angle between I and the x-axis is equal to $\tan^{-1}(I_y/I_x)$, which, in the special case of $A = B$ is $\pi/4$ at all points. The local angle of the reactive component J is equal to $\tan^{-1}(J_y/J_x) = -\pi/4$, irrespective of the ratio A/B.

A more complicated intensity field arises when a plane progressive wave intersects a pure standing wave field. Consider, for example, the case with

$$p_2(x, y, t) = 2B\cos(ky)\exp(i\omega t)$$

Then

$$C_x = (1/2\rho_0 c)[A^2 + 2AB\cos(ky)\cos(kx) + i2AB\cos(ky)\sin(kx)] \quad (4.38)$$

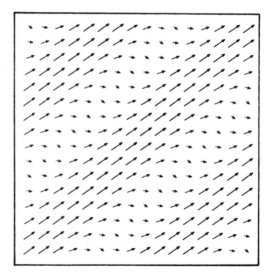

Fig. 4.7. Distribution of mean intensity in the interference field of two orthogonally directed plane progressive waves.[36]

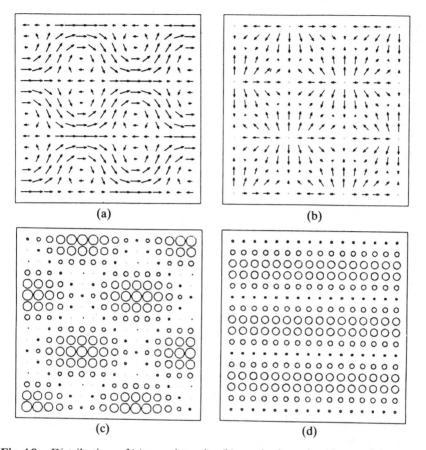

Fig. 4.8. Distributions of (a) mean intensity, (b) reactive intensity, (c) potential energy density and (d) kinetic energy density in the interference field of a plane progressive wave and an orthogonally directed plane standing wave.[36]

and

$$C_y = (1/2\rho_0 c)[2AB\sin(kx)\sin(ky) + i(2AB\cos(kx)\sin(ky) + 2B^2\sin(2ky))]$$

$$(4.39)$$

The distributions of mean and reactive components of intensity, together with those of the potential and kinetic energy densities, are shown in Figs 4.8(a–d). A dramatic difference is observed between Figs 4.7 and 4.8, characterised by the appearance of regions of apparently circulatory energy flow, surrounding points of zero pressure. The reactive intensity vector distribution exhibits regions of divergence, centred on regions of maximum acoustic pressure, and convergence centred on regions of zero pressure; unlike I, the reactive intensity shows no tendency to trace out serpentine paths. Spatial integration of I_x over a wavelength interval of y again yields the mean power in the travelling wave; the corresponding integration over x produces zero.

4.6.2　Characteristics of complex vector intensity

It seems reasonable to assume that, in the regions of Fig. 4.8(a) where the mean intensity vectors form closed circular patterns, there is an active flow of energy around closed circuits: this physical interpretation may, however, be misleading (see Section 4.7).[37,38] Rotation of a field vector is quantified mathematically by the 'curl' of the field; this is defined as the vector product of the operator ∇ with the field vector. In rectangular Cartesian co-ordinates the curl of a field vector $\mathbf{A} = A_x\mathbf{i} + A_y\mathbf{j} + A_z\mathbf{k}$ is defined as

$$\nabla \times \mathbf{A} = [(\partial/\partial x)\mathbf{i} + (\partial/\partial y)\mathbf{j} + (\partial/\partial z)\mathbf{k}] \times \mathbf{A} \tag{4.40}$$

(Note: the vector products of the unit vectors are defined as follows: $\mathbf{i} \times \mathbf{j} = \mathbf{k}$; $\mathbf{j} \times \mathbf{k} = \mathbf{i}$; $\mathbf{k} \times \mathbf{i} = \mathbf{j}$: reversal of order reverses the sign of the result.) Hence the curl of $\mathbf{A}$ is

$$\nabla \times \mathbf{A} = (\partial A_z/\partial y - \partial A_y/\partial z)\mathbf{i} + (\partial A_x/\partial z - \partial A_z/\partial x)\mathbf{j} + (\partial A_y/\partial x - \partial A_x/\partial y)\mathbf{k} \tag{4.41}$$

In two dimensions (x, y) only the third bracketed term applies. We shall restrict our attention mainly to two-dimensional fields because, in general, intensity distributions in three dimensions are too complex to visualise and illustrate.

A simple physical understanding of the significance of the curl of instantaneous intensity may be gained by expressing $\mathbf{I}(t)$ in terms of pressure and particle velocity components in two dimensions. Omitting the symbol of the time dependence of p, u, v and w we obtain

$$\nabla \times \mathbf{I}(t) = [\partial(pv)/\partial x - \partial(pu)/\partial y]\mathbf{k}$$
$$= [p\{\partial v/\partial x - \partial u/\partial y\} + v(\partial p/\partial x) - u(\partial p/\partial y)]\mathbf{k} \tag{4.42}$$

The term in curly brackets is the magnitude of the curl of particle velocity, called the 'vorticity'. This quantity is zero in a sound field in a fluid which is inviscid, and hence lacks the ability to produce shear stresses. From the inviscid fluid momentum equations (3.12)

$$\partial^2 u/\partial y\partial t = -\rho_0\partial^2 p/\partial x\partial y \quad \text{and} \quad \partial^2 v/\partial x\partial t = -\rho_0\partial^2 p/\partial x\partial y$$

and thus

$$\partial/\partial t(\partial u/\partial y - \partial v/\partial x) = 0 \tag{4.43}$$

Hence, an inviscid fluid which initially possesses zero vorticity cannot gain or lose any. Thus eqn (4.42) reduces to

$$\nabla \times \mathbf{I}(t) = [v(\partial p/\partial x) - u(\partial p/\partial y)]\mathbf{k}$$

Again using eqns (3.12)

$$\nabla \times \mathbf{I}(t) = \rho_0[u(\partial v/\partial t) - v(\partial u/\partial t)]\mathbf{k} \tag{4.44}$$

This expression is general for any two-dimensional sound field: it involves products of velocity components in one direction and accelerations in the orthogonal direction, and, if non-zero, implies curvature of the particle trajectory. By specialising to harmonic time dependence, with $u = U\exp[i(\phi_u + \omega t)]$ and $v = V\exp[i(\phi_v + \omega t)]$, and using the real parts of u, v, $\partial u/\partial t$ and $\partial v/\partial t$ in eqn (4.44), we obtain

$$\nabla \times \mathbf{I}(t) = \rho_0 \omega U V \sin(\phi_u - \phi_v)\mathbf{k} \tag{4.45}$$

which is independent of time and therefore equal to the curl of the mean intensity. By implication, the curl of the reactive intensity component is zero. The physical interpretation of eqn (4.44) is that the particle trajectories are elliptical if $\nabla \times \mathbf{I}(t)$ is non-zero.

The instantaneous intensity vector at a point rotates, but at a non-uniform rate. Readers might like to demonstrate this behaviour for themselves by plotting p, u and v as sinusoidal functions of time with arbitrary relative phase, and graphically constructing $\mathbf{I}(t) = p(t)[u(t)\mathbf{i} + v(t)\mathbf{j}]$.

Having considered the vector product of the operator ∇ with vector intensity, we now return to the divergence, which is the scalar product $\nabla.\mathbf{I}(t)$. In Section 4.3 it was shown that, in a time-stationary field, the divergence of the mean intensity in any region is zero in the absence of local generators or absorbers of mean sound power. It is therefore of interest to investigate the divergence of the reactive component of complex intensity. In a two-dimensional harmonic field represented in terms of complex amplitudes of pressure P, and of particle velocity components U and V,

$$\mathbf{J} = J_x\mathbf{i} + J_y\mathbf{j} = \tfrac{1}{2}\operatorname{Im}\{PU^*\}\mathbf{i} + \tfrac{1}{2}\operatorname{Im}\{PV^*\}\mathbf{j} \tag{4.46}$$

$$\nabla.\mathbf{J} = \partial J_x/\partial x + \partial J_y/\partial y \tag{4.47}$$

Now

$$\partial P/\partial x = -i\omega\rho_0 U \quad \text{and} \quad \partial P/\partial y = -i\omega\rho_0 V$$

Hence

$$\operatorname{Im}\{PU^*\} = -(1/\omega\rho_0)\operatorname{Re}\{P(\partial P/\partial x)^*\}$$

and

$$\partial/\partial x[\operatorname{Im}(PU^*)] = -(1/\omega\rho_0)[(\partial P/\partial x)(\partial P/\partial x)^* + \operatorname{Re}\{P(\partial^2 P/\partial x^2)^*\}]$$

with corresponding expressions for the functions of v. Hence

$$\partial J_x/\partial x + \partial J_y/\partial y = -(1/2\omega\rho_0)\{|\partial P/\partial x|^2 + |\partial P/\partial y|^2 \\ + \operatorname{Re}\{P[(\partial^2 P/\partial x^2)^* + (\partial^2 P/\partial y^2)^*]\}\}$$

The Helmholtz equation (3.13b) gives

$$(\partial^2 P/\partial x^2)^* + (\partial^2 P/\partial y^2)^* = -k^2 P^*$$

and hence

$$\begin{aligned}
\nabla.\mathbf{J} &= -(1/2\omega\rho_0)[\omega^2\rho_0^2(|U|^2 + |V|^2) - k^2|P|^2] \\
&= 2\omega[|P|^2/(4\rho_0 c^2) - \tfrac{1}{4}\rho_0(|U|^2 + |V|^2)] \\
&= 2\omega[e_p - e_k] = -2\omega L \tag{4.48}
\end{aligned}$$

in which e_p and e_k are respectively the mean potential and kinetic energy densities, and L is known as the Lagrangian of the field. One physical interpretation of eqn (4.48) is that a local difference between mean potential and kinetic energy densities can be likened to a 'source' of reactive intensity: I consider this interpretation to have little physical justification (see also Ref. 90).

We may summarise the equations satisfied by the complex intensity in a two-dimensional, harmonic field thus:

$$\nabla.\mathbf{I} = 0 \tag{4.49a}$$

$$\nabla \times \mathbf{I} = \rho_0 \omega U V \sin(\phi_u - \phi_v)\mathbf{k} \tag{4.49b}$$

$$\nabla.\mathbf{J} = 2\omega(e_p - e_k) \tag{4.49c}$$

$$\nabla \times \mathbf{J} = 0 \tag{4.49d}$$

Since $\nabla \times \mathbf{J} = 0$, the direction of $\mathbf{J}$ relative to the x-axis is given by $\tan^{-1}(J_y/J_x)$. These equations show that the active component of an intensity field is rotational and the reactive component is solenoidal.[38]

As emphasised by Mann *et al.*[37] and Uosukainen,[38] the process of energy transport in a sound field is described by the temporal evolution of the instantaneous intensity vector field and not by the streamlines corresponding to the mean intensity vectors. Uosukainen says: 'The time average term shows only the temporal net effect of the acoustic energy flow, not what is really happening in the field.' He continues: 'In the case of linear [particle velocity] polarization the energy propagates through a point like "sawing" [see Fig. 4.3]: in the case of elliptical polarization $[\sin(\phi_u - \phi_v) \neq 0$ in eqn (4.45): see also Section 3.7.3] it propagates...making loops in space.' In relation to the initiation and decay of intensity vortices, these authors agree that 'there is no transient trapping of energy but it is always propagating through the vortex region'.

Equation (4.36b) indicates that the mean intensity vector in a harmonic field is directed normal to the surfaces of uniform phase (which are defined as 'wavefronts'); eqn (4.36c) indicates that the reactive intensity vector is directed normal to the surfaces of uniform mean square pressure (or sound pressure level) (Fig. 4.9). In a general harmonic interference field the wavefronts do not propagate at the speed of sound, i.e. the phase speed is not c, as illustrated by the example of a two-dimensional duct in Sections 3.7.3 and 4.8.4: the axial phase speeds of all modes except the plane wave are greater than the speed of sound. This is because a waveguide mode is no more than an interference pattern which happens to match the boundary conditions imposed by the duct walls; the elementary component disturbances which interfere to form this pattern must, of course, propagate at the local speed of sound. The speed of axial propagation of sound energy, the group speed, cannot exceed the local speed of sound, and the time-average axial group speed of energy transport by these modes is less than the speed of sound.

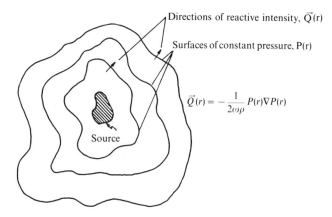

Directions of reactive intensity, $\vec{Q}(r)$

Surfaces of constant pressure, $P(r)$

$$\vec{Q}(r) = -\frac{1}{2\omega\rho} P(r)\nabla P(r)$$

Source

Fig. 4.9. The reactive intensity vector is directed normal to the surfaces of uniform mean square pressure.[37]

The problematic issue of the significance of a mean rate of energy transport is illustrated by the elementary case of the one-dimensional plane wave interference field analysis in Section 4.4. The mean intensity, which is obviously independent of position, is given by eqn (4.24) as

$$\bar{I}_a = (A^2/2\rho_0 c)(1 - R^2)$$

and the mean energy density, which is also independent of position, is given by

$$\bar{e} = (A^2/2\rho_0 c^2)(1 + R^2)$$

Hence one could define a mean speed of energy transport as the ratio of mean intensity to mean energy density, giving

$$\bar{c}_g = c(1 - R^2)/(1 + R^2) \qquad (4.50)$$

This 'speed' varies from the free wave group speed c when the reflection coefficient $R = 0$, to zero when $R = 1$. This is the result of the increase in the proportion of reversed energy flux during part of the cycle as the reactivity of the field increases (see Fig. 4.5).

In view of the complexity of reactive intensity fields created by interference in two- and three-dimensional sound fields it is not surprising that mean intensity streamlines in fields of higher spatial dimensions provide no information about the process of energy transport, as illustrated by consideration of the sound field in a two-dimensional waveguide (Section 4.8.4). The circulatory mean intensity pattern shown in Fig. 4.15(a) is the result of interference between the propagating $n = 0$ plane wave mode and the evanescent $n = 1$ mode. The corresponding instantaneous intensity distribution at intervals of $\frac{1}{16}$th period are shown in Fig. 4.10 which was generated by acoustic finite element analysis.[39] It can be clearly seen that the mean intensity distribution reveals little of the temporal evolution of the energy transport process.

t = T/16

(a)

t = T/8

(b)

t = 3T/16

(c)

t = T/4

(d)

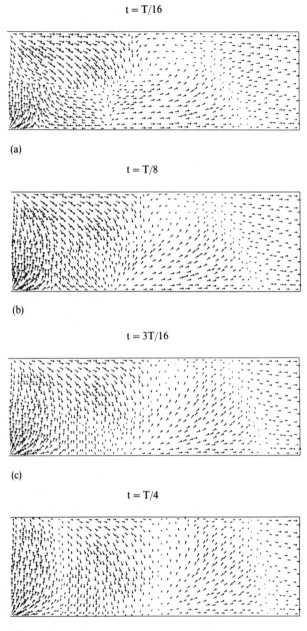

Fig. 4.10. Instantaneous intensity distributions in a two-dimensional duct at intervals of 1/16th period.[39]

t = 5T/16

t = 3T/8

t = 7T/16

t = T/2

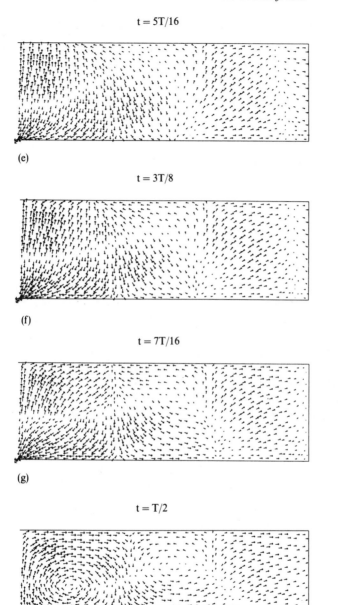

(e)

(f)

(g)

(h)

Fig. 4.10—*contd.*

4.6.3   Vector separation of complex intensity

The complex intensity vector $\mathbf{C} = \mathbf{I} + i\mathbf{J}$ comprises a rotational component, corresponding to mean active intensity, and a solenoidal component corresponding to reactive intensity. The former may be decomposed into one component associated with a scalar potential and another associated with a vector potential,[36] thus:

$$\mathbf{C} = \nabla\beta + \nabla \times \psi + \nabla\gamma \qquad (4.51)$$

The first term has zero curl and the scalar potential β satisfies the Laplace equation

$$\nabla^2\beta = 0 \qquad (4.52)$$

everywhere except in regions of the fluid where active sources or sinks (sound absorbers) operate, in which case

$$\nabla^2\beta = \text{source strength density} \qquad (4.53)$$

Except in linearly polarised fields, the second term in eqn (4.51) has non-zero curl which equals the curl of $\mathbf{I}$. The divergence of both terms is zero. Pascal[36] interprets the scalar potential gradient $\nabla\beta$ as representing the net transfer of sound energy between points in a sound field, with the vector potential component $\nabla \times \psi$ corresponding to interference effects which produce local 'distortions' in the energy flux streamlines, together with apparent mean energy 'circulation', but which involves zero net power flux through any closed surface, even if an active source or absorber is enclosed. This interpretation is disputed by Mann et al.[37] on the basis of a example of an array of three monopoles. It would appear that Pascal's interpretation may not be fully justified, but the question remains open at this time.

4.6.4   Intensity and mean square pressure

Equation (4.36b) gives the ratio of the *r-directed component* of the mean intensity to the local mean square pressure as

$$I_r/\overline{p^2} = -(\partial\phi_p/\partial r)/\omega\rho_0 \qquad (4.54)$$

This relationship forms the basis of derivation of the primary index of quality required of an intensity measurement system in relation to the nature of the field being measured: it is known as the 'Pressure-Intensity Index', symbolised by δ_{pI}.

$$\delta_{pI} = -10\lg[I_r/(\overline{p_0^2}/\rho_0 c)] + 10\lg[\overline{p^2}/p_0^2]\mathrm{dB}$$

$$= 10\lg[\overline{p^2}/I_r] - 10\lg[\rho_0 c] \qquad (4.55a)$$

Hence

$$\delta_{pI} = -10\lg[|\partial\phi_r/\partial r|/k] \qquad (4.55b)$$

δ_{pI}, *as measured*, is a function of both the *form of field* and of the *orientation of the intensity probe axis* within that field.

4.7 INSTANTANEOUS AND TIME-AVERAGED ACTIVE AND REACTIVE INTENSITY

Up to this point in the book, the concept of complex intensity has been presented on the basis of single frequency (pure-tone) models. In practice, sound fields rarely have such simple time dependence, and it is therefore of interest to consider whether the concepts of active and reactive intensity components are tenable for sound fields of arbitrary time-dependence. Jacobsen[40] addresses this question in a paper from which the title of this section was taken.

As shown in Section 4.3, the instantaneous intensity vector $\mathbf{I}(t)$ is defined as the product of instantaneous sound pressure $p(t)$ and the instantaneous sound particle velocity vector $\mathbf{u}(t)$. The time-averaged value of $\mathbf{I}(t)$, which is, in principle, only significant in a time-stationary field, is the active intensity $\mathbf{I}$, irrespective of the specific time dependence. In Fig. 4.5(a), the direction of flow of sound energy does not reverse; the field is purely active. As seen in Figs 4.5(b) and (c) and eqns (4.23) and (4.25), the direction of the instantaneous intensity at any field point can fluctuate with time. Such fluctuation indicates that energy is being transported to and fro, which suggests that a complete characterisation of the form of the intensity field requires a description of the strength of this fluctuation, as a complement to the value of the active intensity. We have seen in earlier sections that the imaginary component $\mathbf{J}$ of a complex intensity $\mathbf{C}$ performs this function is a pure-tone field; but this complex representation is not appropriate to a field having stationary, aperiodic time dependence, in which the fluctuations may be very irregular.

Jacobsen defines a time-averaged reactive sound intensity as

$$\mathbf{J} = \overline{\hat{p}(t)\mathbf{u}(t)}$$

in which $\hat{p}(t)$ is the Hilbert transform of $p(t)$. (The Hilbert transform of a function $f(t)$ is defined as

$$\hat{p}(\tau) = \frac{1}{\pi}\int_{-\infty}^{\infty}\frac{p(t)}{t-\tau}\,\mathrm{d}t\Bigg)$$

In a pure-tone sound field $\mathbf{J}$ is given by eqn (4.36d) as

$$\mathbf{J} = -\nabla\overline{(p^2(t))}/(2\rho c k) \tag{4.56}$$

In a non-pure-tone sound field, one cannot identify separate components of the instantaneous particle velocity which are in phase and in quadrature with the pressure. Various expressions for a complex instantaneous intensity are

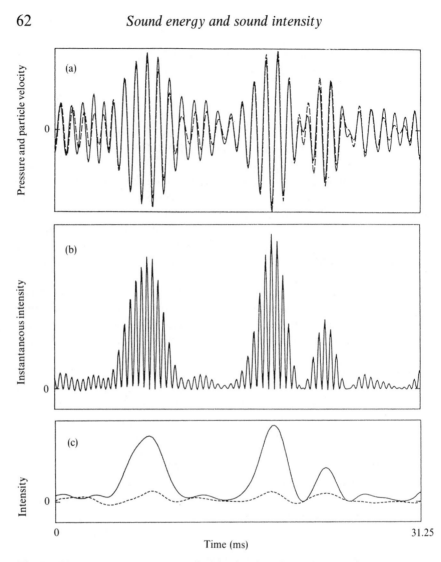

Fig. 4.11(a). Measurements 30 cm from a loudspeaker: (a)—sound pressure, ———
normalised particle velocity; (b) instantaneous intensity; (c) complex instantaneous
intensity:——— real part, ——— imaginary part. One-third octave with centre frequency of
1 kHz.[40]

quoted in Ref. 40, of which the following, due to Heyser,[41] is favoured by
Jacobsen:

$$\mathbf{I}_c(t) = \tfrac{1}{2}(p(t) + \mathrm{j}\hat{p}(t))(\mathbf{u}(t) - \mathrm{j}\hat{\mathbf{u}}(t)) \tag{4.57}$$

This function is a generalisation of the tonal expressions in eqns (4.34a) and
(4.35). In a time-stationary sound field the time-average of $\mathbf{I}_c(t)$ is equal to
$\mathbf{I} + \mathrm{i}\mathbf{J}$. In a narrow-band sound field the real and imaginary parts of $\mathbf{I}_c(t)$ are
low-pass signals which represent the running short-term averages of the active

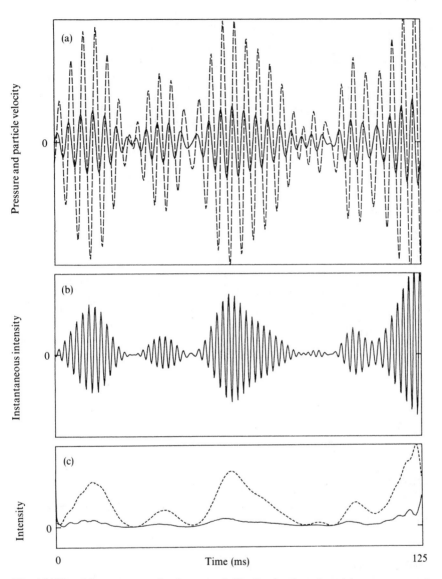

Fig. 4.11(b). Measurements in the near field of a loudspeaker: (a) ———— sound pressure, ––– normalised particle velocity; (b) instantaneous intensity; (c) complex instantaneous intensity: ———— real part, –––– imaginary part. One-third octave band with centre frequency of 250 Hz.[40]

and reactive components of the complex instantaneous intensity. Measurements of these quantities under various field conditions are presented in Figs 4.11(a–d). The real part of the complex instantaneous intensity is seen to indicate the magnitude of local mean energy flow, whereas the imaginary part

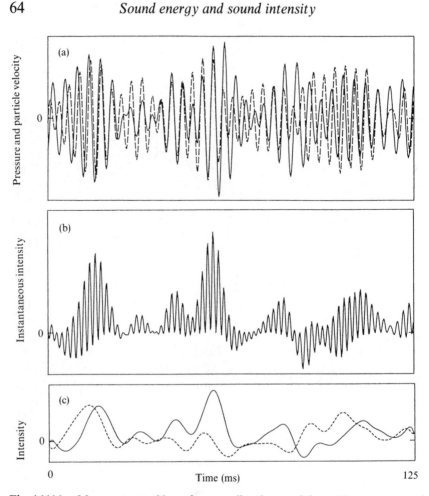

Fig. 4.11(c).   Measurements 30 cm from a vibrating steel box: (a) ——— sound pressure, – – – normalised particle velocity; (b) instantaneous intensity; (c) complex instantaneous intensity: ——— real part, – – – imaginary part. One-third octave band with centre frequency of 250 Hz.[40]

peaks during periods when the rapid and strong reversals of energy flow occur. Figs 4.11(b and d) reveal why it is necessary to average for such a long time to estimate the narrow-band mean intensity correctly in the near fields of radiators and in reverberant fields. Although the significance of the components of the complex instantaneous intensity decreases as the bandwidth of the field increases, it should be remembered that the process of measuring intensity in a finite band is equivalent to restricting the source bandwidth to that band, however broad its total bandwidth. Consequently, the associated imaginary component of the complex instantaneous intensity forms a useful indicator of the averaging time necessary to yield a good estimate of the mean intensity.

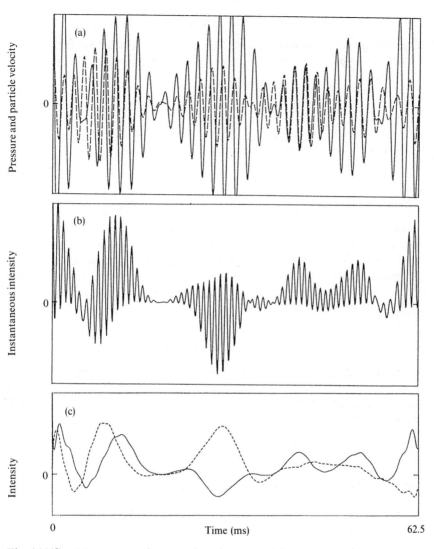

Fig. 4.11(d). Measurement in a reverberation room: (a) ———— sound pressure, ———
normalised particle velocity; (b) instantaneous intensity; (c) complex instantaneous
intensity: ———— real part, ——— imaginary part. One-third octave band with centre
frequency 500 Hz.[40]

4.8 EXAMPLES OF IDEALISED SOUND INTENSITY FIELDS

The following examples are presented in order to illustrate the various
characteristics of harmonic intensity fields revealed by the aforegoing analyses.

4.8.1 The point monopole

Expressions for the pressure and radial particle velocity fields, presented earlier in Chapter 3, are repeated here for the convenience of the reader.

$$p(r, t) = (A/r)\exp[i(\omega t - kr)]$$

$$u_r(r, t) = (A/\omega \rho_0 r)(k - i/r)\exp[i(\omega t - kr)]$$

Application of eqns (4.16) and (4.18) yields the following expressions for the active and reactive intensity components, which are purely radially directed:

$$I_a(r, t) = (A^2/2r^2\rho_0 c)[1 + \cos 2(\omega t - kr)] \tag{4.58}$$

and

$$I_{re}(r, t) = (A^2/2r^3\rho_0 \omega) \sin 2(\omega t - kr) \tag{4.59}$$

The ratio of the magnitudes $|I_a|/|I_{re}| = |I/J| = kr$, which shows that the reactive intensity dominates in the near field, and the active component dominates in the far field. The relationship between I and p^2 is the same as in a plane progressive wave. The curl of $\mathbf{I}$ is zero because the intensity is directed purely radially. The Lagrangian equals $A^2/\omega \rho_0 r^4$, indicating that the divergence of $\mathbf{J}$ decreases very rapidly with r.

4.8.2 The compact dipole

The ideal dipole source illustrated in Fig. 4.12 comprises two point monopoles of equal strength and opposite polarity in close proximity to each other in terms of a wavelength ($kd \ll 1$). The pressure field of a harmonic dipole is expressed by

$$p(r_1, r_2, t) = [i\omega \rho_0 Q/4\pi][(1/r_1)\exp(-ikr_1) - (1/r_2)\exp(-ikr_2)]\exp(i\omega t) \tag{4.60}$$

In terms of the general expression $p = P\exp(i\phi_p)\exp(i\omega t)$

$$P = (\omega \rho_0 Q/4\pi)[1/r_1^2 + 1/r_2^2 - (2/r_1 r_2)\cos k(r_1 - r_2)]^{1/2} \tag{4.61}$$

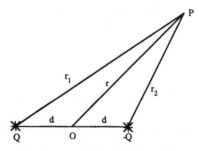

Fig. 4.12. Compact dipole of strength $2i\omega Qd$.

and

$$\phi_p = \tan^{-1}[(r_2 \cos kr_1 - r_1 \cos kr_2)/(r_2 \sin kr_1 - r_1 \sin kr_2)] \quad (4.62)$$

Expressions for radial and tangential components of active and reactive intensity may be derived by the application of eqns (4.16) and (4.18) in the appropriate directions. Alternatively, eqns (4.34c) and (4.34d) may be employed with the radial component of complex intensity being given by

$$C_r = I_r + iJ_r = \tfrac{1}{2}(PU_r^*) \quad (4.63a)$$

and the tangential component as

$$C_\theta = I_\theta + iJ_\theta = \tfrac{1}{2}(PU_\theta^*) \quad (4.63b)$$

The complex pressure amplitude may be written as

$$P = (P_{1r} + P_{2r}) + i(P_{1i} + P_{2i}) \quad (4.64)$$

where

$$P_{1r} = (\omega\rho_0 Q \sin(kr_1))/(4\pi r_1)$$
$$P_{2r} = -(\omega\rho_0 Q \sin(kr_2))/(4\pi r_2)$$
$$P_{1i} = (\omega\rho_0 Q \cos(kr_1))/(4\pi r_1)$$
$$P_{2i} = -(\omega\rho_0 Q \cos(kr_2))/(4\pi r_2)$$

The complex amplitude of the radial velocity component is given by

$$U_r = U_{rr} + iU_{ri}$$

where

$$U_{rr} = (1/\omega\rho_0)\{(P_{1i}/r_1 + kP_{1r})(1/r_1)(r + d\cos\theta)$$
$$+ (P_{2i}/r_2 + kP_{2r})(1/r_2)(r - d\cos\theta)\}$$
$$U_{ri} = (1/\omega\rho_0)\{(-P_{1r}/r_1 + kP_{1i})(1/r_1)(r + d\cos\theta)$$
$$+ (-P_{2r}/r_2 + kP_{2i})(1/r_2)(r - d\cos\theta)\} \quad (4.65)$$

The complex amplitude of tangential velocity is given by

$$U_\theta = U_{\theta r} + iU_{\theta i}$$

where

$$U_{\theta r} = (1/r\omega\rho_0)\{(P_{1i}/r_1 + kP_{1r})(-d\sin\theta/r_1)$$
$$+ (P_{2i}/r_2 + kP_{2r})(d\sin\theta/r_2)\}$$
$$U_{\theta i} = (1/r\omega\rho_0)\{(-P_{1r}/r_1 + kP_{1i})(-d\sin\theta/r_1)$$
$$+ (-P_{2r}/r_2 + kP_{2i})(d\sin\theta/r_2)\} \quad (4.66)$$

The resulting expressions for I and J can be decomposed into components corresponding to the two individual monopoles, plus 'interference' terms which introduce tangential components absent in the individual monopole fields. Examples of the distributions of I and J are presented in Figs 4.13(a, b).

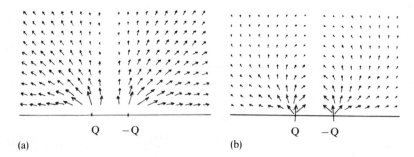

Fig. 4.13. Distributions of (a) mean intensity and (b) reactive intensity in the field of a compact dipole ($kd = 0\cdot2$; vector scales $\propto I^{1/4}$, $J^{1/4}$: (a) scale 16 times (b) scale).

4.8.3 Interfering monopoles

As mentioned in Chapter 3, it is possible mathematically to synthesise any complex source from an array of point monopoles of suitable amplitude and phase. The form of the intensity field produced by the superposition of such elementary wave fields may be illustrated by the case of two point monopoles of variable relative amplitude and phase. Figure 4.14(a) illustrates the mean intensity field of monopoles of equal strength and phase at a non-dimensional separation distance of $kd = 0\cdot2$ (the plotted vector magnitudes are propor-

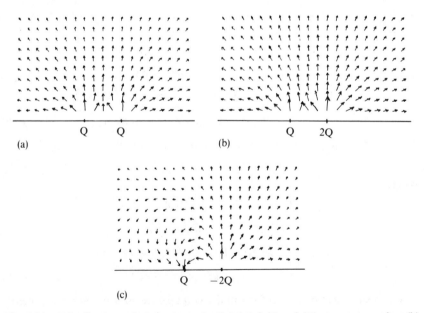

Fig. 4.14. Distribution of mean intensity in the fields of (a) two monopoles, (b) monopoles of different strengths but equal phase and (c) monopoles of different strengths and opposite phase.

tional to $I^{1/4}$). The effect of doubling the strength of one of the sources is illustrated in Fig. 4.14(b), and Fig. 4.14(c) shows what happens when the phase of one of the pair is reversed (scales $\frac{1}{16}$ relative to Fig. 4.14(a)). The result clearly demonstrates the fact that the power radiated by an elementary volumetric source is affected by the pressure field imposed upon it by other coherent (phase-related) sources in its proximity. The magnitude of the mutual influence depends upon the separation distance because of the inverse square law. In the last case the weaker monopole constitutes an active *sink*. (The phenomenon can easily be demonstrated with two small loudspeakers and an intensity meter.)

These results demonstrate an interdependence between coherent source regions which, in principle, makes it impossible, and even illogical, to identify any one region as *the source* of sound power, because the total power is a consequence of the simultaneous action of all the regions. An implication of considerable import for the practice of noise control is that the suppression of any portion of a total source array does not necessarily reduce the radiated power; indeed, it may increase it. A related consequence is that measurements of intensity in regions close to an extended source may not be well suited to the task of estimating the total radiated power because they are 'contaminated' by components of active energy flow which do not leave the vicinity of the source.

In case the reader is about to cast away this 'unhelpful' volume in frustration or despair, it should be pointed out that traverses of intensity probes over the surfaces of real sources often produce useful information. Among the reasons for this apparent contradiction of the above strictures are the following: (i) interdependence (in technical terminology 'mutual radiation impedance') exists only between coherently fluctuating source regions, i.e. those having time-stable, unique phase relationships; (ii) in cases of broad band sources, a multi-frequency 'smearing effect', illustrated in the following section, operates so as to reduce the degree and extent of near field recirculation of energy which is characteristic of narrow frequency bands.

4.8.4   Intensity in ducts

As shown in Chapter 3, acoustic fields within uniform ducts may be considered to consist of summations of the characteristic duct modes, the amplitudes and phases of which are determined by the forms and locations of the sources driving the fields. These modes are simply interference patterns which satisfy the particular duct boundary conditions. Equation (3.35) indicates that the axial wavenumber components of modes other than the plane wave (the higher order modes) are, at frequencies higher than their cut-off frequencies, greater than the acoustic wavenumber $k = \omega/c$. At first sight, the implication appears to be that sound can propagate along the duct axis at speeds higher than c: this is, however, not so, because it is not the acoustic disturbances forming the modal interference pattern which are propagating at this speed, but the pattern itself. The group speed at which the modal *energy* propagates (given by $c_g = \partial\omega/\partial k_n$) is less than c at all frequencies.

The intensity distribution in an *isolated* mode of a uniform planar duct of width a, with rigid walls, is obtained from the expressions for modal pressure and component particle velocities given by eqns (3.36)–(3.38). The axial component of mean intensity is

$$I_{nx} (1/2\omega\rho_0) \operatorname{Re}\{k_n\} [P_n \cos(n\pi y/a)]^2 \tag{4.67}$$

which is non-zero only at frequencies above the mode cut-off frequency, when k_n is real (see eqn (3.35)). The transverse component of mean intensity I_{ny} is zero at all frequencies. The power per unit depth propagating along the duct above cut-off is given by

$$W_{nx} = \int_0^a I_x \mathrm{d}y = (1/4\omega\rho_0) k_n P_n^2 \tag{4.68}$$

For a given modal pressure amplitude, the power varies from zero, at cut-off, to half the plane wave value at infinite frequency.

Sources in ducts simultaneously excite a number of modes; at any one frequency, a proportion of these propagate, and the rest do not. The associated intensity distribution is produced by the action of the sum of the modal pressures on the sum of the modal particle velocities, thus:

$$I_x = (1/2\omega\rho_0) \operatorname{Re}\left\{\left[\sum_{n=0}^{\infty} P_n \cos(n\pi y/a) \exp(-\mathrm{i}k_n x)\right]\right.$$
$$\left. \times \left[\sum_{m=0}^{\infty} k_m P_m^* \cos(m\pi y/a) \exp(\mathrm{i}k_m x)\right]\right\} \tag{4.69}$$

and

$$I_y = -(1/2\omega\rho_0) \operatorname{Im}\left\{\left[\sum_{n=0}^{\infty} P_n \cos(n\pi y/a) \exp(-\mathrm{i}k_n x)\right]\right.$$
$$\left. \times \left[\sum_{m=0}^{\infty} (m\pi/a) P_m^* \sin(m\pi y/a) \exp(\mathrm{i}k_m x)\right]\right\} \tag{4.70}$$

This distribution is extremely complicated because pressures in non-propagating (cut-off) modes co-operate with the particle velocities in propagating modes, and vice versa. Non-propagating modes contribute significantly to the total intensity field in the vicinity of sources, and in regions of change of duct properties such as cross-sectional area, or wall impedance: however, they do not contribute to the total power transmitted through any cross-section (unless the duct is very short and has a reflective termination).

Some theoretical examples of mean intensity distributions in an infinitely long (or anechoically terminated) two-dimensional, uniform duct with rigid walls, excited by a point monopole source, are presented in Figs 4.15(a–f). The circulatory flow pattern seen in Fig. 4.15(a) results from 'co-operation' between the propagating plane wave ($n = 0$) and the non-propagating $n = 1$

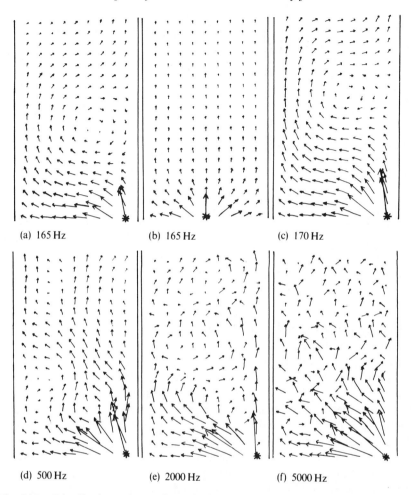

(a) 165 Hz   (b) 165 Hz   (c) 170 Hz

(d) 500 Hz   (e) 2000 Hz   (f) 5000 Hz

Fig. 4.15. Distributions of mean intensity in a planar infinite duct excited by a point monopole source (duct width = 1 m, $f_{10} = 170$ Hz; vector scale $\propto I^{1/4}$).

mode, which has a cut-off frequency of 170 Hz; it disappears when the source is moved to the nodal plane of the latter (Fig. 4.15(b)). In Fig. 4.15(c) both the $n = 0$ and $n = 1$ modes propagate, and the circulatory pattern actually repeats indefinitely along the duct length.

Some examples of high frequency, multi-mode intensity distributions in the same duct are shown in Figs 4.15(d–f). Figures 4.16(a–c) show distributions produced by superimposing the intensity fields at a number of frequencies uniformly distributed within a frequency band; these approximate to those which would be generated by band-limited white noise. These latter are seen to be less complex than the single frequency patterns, exhibiting much less tendency to form circulatory cells. This 'smearing' effect may be likened to that which occurs in the distribution of mean square pressure in reverberant

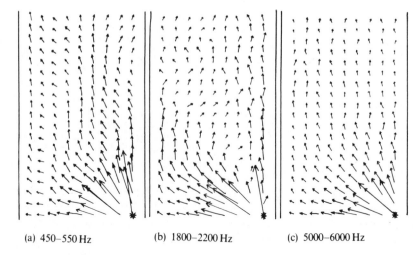

(a) 450–550 Hz   (b) 1800–2200 Hz   (c) 5000–6000 Hz

Fig. 4.16. Distribution of mean intensity in a planar infinite duct excited by a multi-frequency point monopole source (vector scale $\propto I^{1/4}$).

enclosures excited by broad band sources. One consequence of this behaviour is that excessively narrow frequency resolution of broad band intensity fields is likely simply to confuse the observer; one-third octave bands are usually sufficiently narrow to reveal frequency trends. A corollary is that over-reliance on the results of theoretical single frequency analysis of intensity fields in enclosures may give grounds for unjustified pessimism in relation to the spatial sampling requirements, and the interpretation of enclosed fields in general, except when the source is harmonic.

The intensity distributions shown in Figs 4.15 and 4.16 were actually computed using a line array of image sources, as described in Section 3.7.3, and not by modal summation. In the region of a duct close to the source plane there exists a strong near field, which would require a large number of cut-off modes accurately to represent it; in contrast, only a few image sources are required. In regions of a duct remote from the source plane, the modal representation is more efficient, because only the limited set of propagating modes needs to be included.

4.8.5   Sound fields in enclosures: the diffuse field

Sound fields which result from the multiple reflection of sound waves by the boundaries of an enclosure are extremely complex in form and generally not amenable to detailed and precise quantitative description. Schultz[42] expressed his wonder as follows: '... it is almost incredible to me that we could produce such a complex and mysterious thing, just by putting up four walls, a floor and a ceiling, and then radiating sound into it. And yet the more we study sound in an enclosed space, the more peculiar it seems.'

In order to develop analyses of bounded sound fields, theoreticians assume idealised models which are abstractions from physical reality. The deductions and conclusions from analysis of such models are only valid if they correspond closely with physical reality in those features which significantly influence the characteristics and behaviour under examination. Any assumptions should be thoroughly checked against physical observation in order to establish their ranges of applicability. Perhaps the most controversial model is that of the 'diffuse' field. When sound is generated within a space having rather reflective boundaries, the field at any point comprises the superposition of many reflections of the sound wave originating from the source(s); the associated wavevectors obviously point in a multitude of directions. If the source is time-harmonic (pure tone), interference between the intersecting waves produces a stationary wave pattern which exhibits large spatial variations of mean square pressure and associated acoustic quantities. If the reflectivity of the boundaries is not greatly frequency-dependent, the degree of spatial variation is largely independent of source frequency; but the particular spatial distribution varies strongly with frequency. Consequently, if the source bandwidth is increased, the spatial distributions of these field quantities 'smear out' to produce less spatial variation, and the associated directional distribution of the mean intensity vectors follows suit. This behaviour constitutes one good reason why it is asking for trouble to measure the intensity of broad band fields in a reverberant enclosure in very narrow bands if it can be avoided. In the limit of a large frequency bandwidth one could imagine that the variation becomes negligible, provided that the field does not contain, within the bandwidth, dominant resonant acoustic mode components at particular frequencies which would impose their particular spatial patterns on the field.

One model of a diffuse field represents the field at any point in space as receiving contributions from uncorrelated plane waves having uniformly probable wavevector direction. Such a sound field could, in principle, be generated in a small volume of free space by a large number of small, uncorrelated sound sources distributed uniformly over a spherical surface centred on that point. It is clear that a corresponding physical pure-tone 'diffuse' field cannot exist, since pure-tone sources and their images in reflecting boundaries are fully correlated. However, for the purposes of mathematical analysis, it may be reasonable to assume that all possible directions of the particle velocity vector, and all possible phases relative to scalar pressure, are uniformly probable in a pure-tone sound field in a large reverberant room, even if the field quantities themselves vary considerably from place to place. This assumption necessarily neglects the presence of the direct field of the source and of any net flux of energy from the source to the regions of sound absorption which can only be explained in terms of some underlying systematic structure to the sound field. Clearly, the diffuse field model is increasingly realised as the bandwidth of the source increases.

The time-average intensity is equal to the zero-time-delay cross-correlation between instantaneous pressure and particle velocity (see eqn (5.9)). The degree of zero-time-delay correlation at any field point between the waves radiated directly by a source and the individual reflections of those waves, is a function of source (or analysis) bandwidth and the time difference between the arrivals of the two components. Many surfaces produce reflected waves which are relatively undistorted versions of the incident wave, albeit attenuated. Consequently, the cross-correlation function between the two closely approximates the autocorrelation function of the latter. If the time delay between arrivals significantly exceeds the decay time of the autocorrelation function, the zero-time-delay cross-correlation will be zero. The decay time is inversely proportional to signal bandwidth: hence, broad band incident and reflected waves will be largely uncorrelated, except very close to a reflecting surface, which means that the time-average intensity field contains no interference terms, and the individual time-average intensity vectors may be added. This is why the multi-frequency intensity distributions shown in Fig. 4.16 are much simpler in form than those for a single frequency. Remarkably, all reflected waves, together with the directly radiated wave, are fully *coherent*, even in a very reverberant space, provided that they are generated by a coherent (single) source, because they have a common origin. This apparent paradox is resolved once one realises that the zero-time-delay cross-correlation function is given by the integral over frequency of the even part of the cross-spectrum (see eqns (5.11) and (5.12)). The individual frequency components of the cross-spectrum may be either positive or negative, and hence the integral may be vanishingly small, even when these components are large and non-zero. This distinction between 'correlation' and 'coherence' is very important, but often not clearly made.

The statistical properties of random wave fields have been extensively studied; they are the subject of comprehensive reviews by Jacobsen[43] and Ebeling.[44] Ebeling's model represents a *pure-tone* sound field which is assumed to comprise a large number of elementary waves resulting from some scattering or multiple reflection process. The essential assumption is that the amplitudes and phases of the component waves contributing to the sound field at any point are statistically independent and that the phases are uniformly distributed within the interval $-\pi$ to π. Application of the central limit theorem leads directly to expressions for the probability densities and variances of the real and imaginary parts of the complex amplitude of the total field (which are equal). The real and imaginary parts are uncorrelated. A result of particular interest is that the mean energy flux (active intensity) is proportional to the imaginary part of the first-order space derivative of the spatially stationary component of the spatial correlation of the pressure.

This model also yields expressions for the distribution of active intensity in a *pure-tone* random wave field. For a specific direction $\mathbf{r}$, the probability density distribution of the mean intensity component I_r is given by

$$p(I_r) = \frac{1}{(2\langle I_r^2 \rangle - 3\langle I_r \rangle^2)^{1/2}} \times \exp \left\{ \frac{(2\langle I_r^2 \rangle - 3\langle I_r \rangle^2)^{1/2}|I_r|}{\langle I_r^2 \rangle - 2\langle I_r \rangle^2} + \frac{\langle I_r \rangle I_r}{\langle I_r^2 \rangle - 2\langle I_r \rangle^2} \right\}$$

(4.71)

which is, in general, asymmetric on account of net energy flow. In cases when $\langle I_r \rangle = 0$, this reduces to

$$p(I_r) = \frac{1}{(2\langle I_r^2 \rangle)^{1/2}} \exp\left\{ -\frac{\sqrt{2}|I_r|}{(\langle I_r^2 \rangle)^{1/2}} \right\} \qquad (4.72)$$

which indicates that the intensity has a symmetrical Laplacian distribution. The corresponding expressions for the mean intensity in two and three dimensions are exceedingly complicated.

Figure 4.17 shows the distribution of mean intensity near to an absorbing sheet in a reverberation chamber driven by a pure-tone source. The results of measurements of the joint distribution of two orthogonal components of sound intensity in a reverberation room containing diffusors are presented in Figs 4.18(a) and (b). In one plane, the intensity is almost isotropic, whereas in the other it is clearly not so, thereby indicating some net transport of energy in this plane. A comparison of the theoretical distribution of the magnitude of the mean intensity in an isotropic field and results of measurements made in

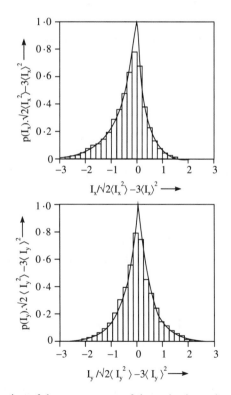

Fig. 4.17. Distribution of the components of the active intensity vector at the face of an absorber with an absorption coefficient of 0·75. The x-component points in the direction perpendicular to the face of the absorber: the y-direction is parallel to the face of the absorber.[44]

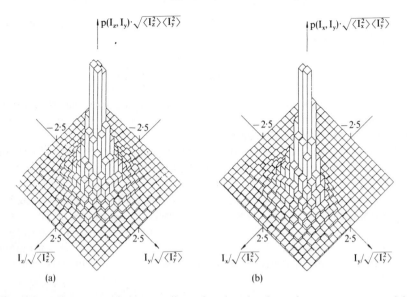

Fig. 4.18. Histogram of the two-dimensional active intensity vector measured in a reverberation room with diffusors. In the y, z plane in (a) the flux is almost isotropic, whereas in the x, y plane in (b) the energy flux is strongly asymmetric.[44]

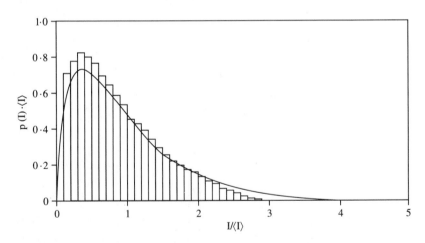

Fig. 4.19. Histogram of the modulus of the two-dimensional active intensity vector for the condition shown in Fig. 4.18(a). The theoretical curve holds for an isotropic sound field.[44]

a reverberation room containing diffusors is presented in Fig. 4.19; some deviation from isotropy is observed.

A somewhat more accessible analysis of intensity distribution in an ideal, *pure-tone* diffuse field is presented by Jacobsen.[45] In his model, which applies to reverberant fields in enclosures of above the Schroeder 'large room' frequency ($f = 2000 \ (T_{60}/V)^{1/2}$), the normalised spatial variances of mean square pressure and mean square particle velocity are unity, and both quantities are exponentially distributed. The basic assumption central to this model is that all phase angles between pressure and any one directed component of particle velocity are equally likely. (This model does not account for the direct field of a source in which the phase between pressure and particle velocity is clearly not random.) This assumption leads to the conclusion that the spatial-average value of active intensity component $\langle I_r \rangle$ is zero. It is important to understand that this conclusion does not imply that I_r is zero everywhere; it simply suggests that the distribution of I_r is symmetric about zero. In fact, the intensity distribution in a pure two-dimensional standing wave field, which takes the form of closed cells of circulating energy flux, also satisfies this condition. It is shown further that the normalised spatial variance of the *magnitude* of any I_r is equal to unity. Since all phase angles ϕ between pressure and particle velocity are assumed to be equally likely, it is obvious that the statistical properties of any reactive intensity component J_r (proportional to sin ϕ) are identical to those of I_r (proportional to cos ϕ), which implies that such a field may be considered to be equally active and reactive. Since I_r has a Laplacian distribution (eqn (4.72)), its magnitude must be exponentially distributed. A conclusion of considerable interest is that, in such a pure-tone diffuse field, the expected difference between the level of the space-average sound pressure and the level of the space-average value of the *magnitude* of any directed component of either active or reactive intensity (a global form of δ_{pI}) will be approximately 5.5 dB. In the case of a room having reverberation time T excited by noise of bandwidth B (Hz), this difference is expected to increase to $5 \lg \ [3\pi(1 + BT_{60}/3 \ln 10)]$. An implication of practical consequence for source location and for sound power determination is that the direct field intensity of a source radiating into a reverberant enclosure will dominate the intensity field within a region having a radius between three and eight times the classical 'reverberation radius'.

4.8.6 The normal velocity wave on the boundary of a semi-infinite fluid

Many sources of sound take the form of vibrating solid surfaces: it is therefore of considerable practical interest to study the intensity fields created by such sources. The simplest idealised model of a vibrating surface is a normal velocity wave imposed on an infinite plane surface which bounds a semi-infinite fluid. We shall initially analyse the intensity fields of two such idealised models before proceeding to the more realistic case of a bounded, vibrating plate in a baffle.

A harmonic wave of normal velocity is supposed to travel in the x-direction along the boundary at $y = 0$ of a semi-infinite region of fluid, with frequency ω, amplitude V_n and wavenumber k_t (Fig. 4.20). The pressure field in the fluid is

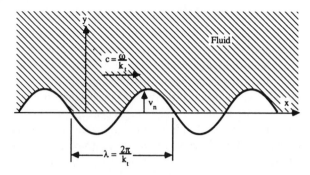

Fig. 4.20. Travelling wave on a fluid space boundary.

given by[33]

$$p(x, y, t) = (\rho_0 \omega V_n/k_y) \exp[i(\omega t - k_t x - k_y y)] \quad k_t < k \qquad (4.73a)$$

where, in order to satisfy the wave equation, $k_y = (k^2 - k_t^2)^{1/2}$ and

$$p(x, y, t) = (i\rho_0 \omega V_n/k_y) \exp(-k_y y) \exp[i(\omega t - k_t x)] \quad k_t > k \qquad (4.73b)$$

where $k_y = (k_t^2 - k^2)^{1/2}$. In the former case, where the wave is supersonic, the expressions for the squared pressure amplitude and the pressure phase are

$$P^2 = (\rho_0 \omega V_n)^2/(k^2 - k_t^2)$$

and

$$\phi_p = -k_t x - k_y y$$

Thus, according to eqns (4.16) and (4.18),

$$I_x = \rho_0 \omega V_n^2 k_t / 2(k^2 - k_t^2) \qquad (4.74a)$$

$$I_y = \rho_0 \omega V_n^2 / 2(k^2 - k_t^2)^{1/2} \qquad (4.74b)$$

$$J_x = J_y = 0 \qquad (4.74c)$$

In this case $\nabla \times \mathbf{I} = 0$. The intensity is purely active, and the energy travels in straight lines at an angle to the x-axis equal to $\tan^{-1}(I_y/I_x) = \cos^{-1}(k_t/k)$.

When $k_t > k$, the wave is subsonic, and the corresponding expressions are

$$P^2 = [\rho_0 \omega V_n^2/(k_t^2 - k^2)] \exp(-2k_y y)$$

$$\phi_p = \pi/2 - k_t x$$

$$I_x = \tfrac{1}{2}[(\rho_0 \omega V_n^2 k_t)/(k_t^2 - k^2)] \exp(-2k_y y) \qquad (4.75a)$$

$$I_y = 0 \qquad (4.75b)$$

$$J_x = 0 \qquad (4.75c)$$

and

$$J_y = \tfrac{1}{2}[\rho_0 \omega V_n^2/(k_t^2 - k^2)^{1/2}] \exp(-2k_y y) \qquad (4.75d)$$

In this case,

$$\nabla \times \mathbf{I} = [\rho_0 \omega V_n^2 k_t/(k_t^2 - k^2)^{1/2}] \exp(-2k_y y)\mathbf{k} \quad (4.75e)$$

$$\nabla.\mathbf{J} = -\rho_0 \omega V_n^2 \exp(-2k_y y) \quad (4.75f)$$

and

$$\nabla.\mathbf{I} = \nabla \times \mathbf{J} = 0$$

as they must.

In this case, there is a mean flow of energy parallel to the surface; this is driven by the surface in the form of a surface wave which decays exponentially with normal distance from the surface. Reactive oscillation of energy, which exhibits similar decay characteristics, occurs purely in a direction normal to the surface, because the mean square pressure does not vary over planes parallel to the surface.

A standing boundary wave may be constructed by superimposing two waves of equal amplitude travelling in opposite directions. In this case

$$p(x, y, t) = [2\rho_0 \omega V_n/(k^2 - k_t^2)^{1/2}] \cos(k_t x) \exp[i(\omega t - k_y y)] \; k_t < k \quad (4.76)$$

and

$$p(x, y, t) = [2i\rho_0 \omega V_n/(k_t^2 - k^2)^{1/2}] \exp(-k_y y) \exp(i\omega t): k_t > k \quad (4.77)$$

In the first case, when the component waves have supersonic phase speeds,

$$P^2(x, y) = 4(\rho_0 \omega V_n)^2 \cos^2(k_t x)/(k^2 - k_t^2)$$

and

$$\phi_p = -k_y y$$

Hence

$$I_x = 0 \quad (4.78a)$$

$$I_y = 2\rho_0 \omega V_n^2 \cos^2(k_t x)/(k^2 - k_t^2)^{1/2} \quad (4.78b)$$

$$J_x = \rho_0 \omega V_n^2 k_t \sin(2k_t x)/(k^2 - k_t^2) \quad (4.78c)$$

and

$$J_y = 0 \quad (4.78d)$$

The curl of the mean intensity is given by

$$\nabla \times \mathbf{I} = -2\rho_0 \omega V_n^2 k_t [\sin(2k_t x)/(k^2 - k_t^2)^{1/2}]\mathbf{k} \quad (4.78e)$$

but the divergence of $\mathbf{I}$ is, of course, zero. The divergence of the reactive intensity is

$$\nabla.\mathbf{J} = 2\rho_0 \omega V_n^2 k_t^2 \cos(2k_t x)/(k^2 - k_t^2) \quad (4.78f)$$

The distribution of $\mathbf{I}$ and $\mathbf{J}$ is shown in Fig. 4.21. Mean energy flow takes place only normal to the surface, and the reactive energy oscillation occurs in planes parallel to the surface. Interestingly, although the curl of the mean intensity is non-zero, no circulatory active energy flow pattern is to be seen.

In the case of subsonic travelling wave components

$$P^2 = 4[(\rho_0 \omega V_n)^2 \cos^2(k_t x)/(k_t^2 - k^2)] \exp(-2k_y y)$$

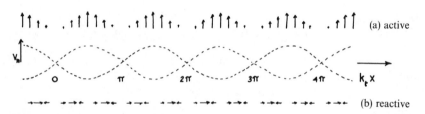

(a) active

(b) reactive

Fig. 4.21. Distributions of (a) mean intensity and (b) reactive intensity in the field of an infinite plate carrying a plane standing wave ($k_1/k = 0.2$).

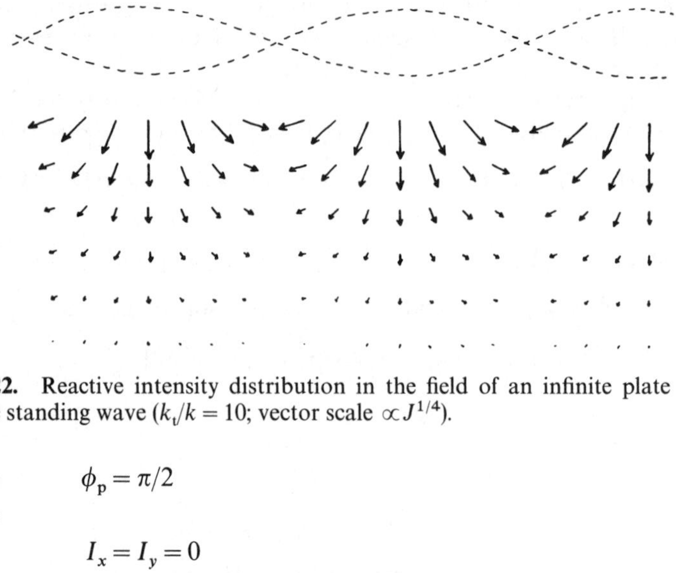

Fig. 4.22. Reactive intensity distribution in the field of an infinite plate carrying a plane standing wave ($k_1/k = 10$; vector scale $\propto J^{1/4}$).

$$\phi_p = \pi/2$$

Hence

$$I_x = I_y = 0 \tag{4.79a}$$

$$J_x = [\rho_0 \omega V_n^2 k_t \sin(2k_t x)/(k_t^2 - k^2)]\exp(-2k_y y) \tag{4.79b}$$

and

$$J_y = 2[\rho_0 \omega V_n^2 \cos^2(k_t x)/(k_t^2 - k^2)^{1/2}]\exp(-2k_y y) \tag{4.79c}$$

This reactive intensity distribution is shown in Fig. 4.22 for $k/k_t = 0.1$, scaled as $J^{1/4}$. There is no active energy flow. The curl of the reactive intensity is, of course, zero and the divergence is given by

$$\mathbf{V.J} = -4\rho_0 \omega V_n^2 \cos^2(k_t x)\exp(-2k_y y)$$

$$+ [2\rho_0 \omega V_n^2 k_t^2 \cos(2k_t x)/(k_t^2 - k^2)]\exp(-k_y y) \tag{4.79d}$$

4.8.7   Radiation from a vibrating surface

The radiation fields of real, bounded structures will exhibit a mixture of the field characteristics displayed by the examples of the previous section. A simple two-dimensional case is analysed here in order to illustrate certain features of import for the practical application of sound intensity measurement of vibrating structures.

A narrow baffled strip of finite length is supposed to vibrate in a sinusoidal pattern, as previously illustrated in Fig. 3.12. Time-dependent intensity distributions in the plane of the strip normal to the baffle are illustrated in Figs 4.23(a–d): note the reversal of intensity vectors during the cycle, indicating the presence of a reactive intensity component. Figures 4.24(a–d) show mean intensity distributions for a range of the frequency parameter $K = k/k_b$, where k_b is the principal vibrational wavenumber of the mode of vibration; in Fig. 4.24(d), the sinusoid has been modulated by an exponential spatial decay, which reduces the reactive energy flow by reducing volumetric cancellation.

The results show that the circulation process in the near field, in which power flows back into the surface in sink regions, weakens as K approaches unity, the wavenumber ratio which divides the ranges of inefficient and efficient radiation.[33]

4.9   THE HELMHOLTZ RESONATOR

A Helmholtz resonator consists of an enclosed volume of air which communicates with its surroundings via a neck. This arrangement constitutes an acoustic resonator which may be employed variously as a frequency-selective sound absorber, a low impedance element in a waveguide and an energy storage device. A properly tuned resonator can extract energy very effectively from a sound field in which it is located, apparently 'sucking in' energy from a surrounding region which greatly exceeds its physical dimensions: this it achieves through a process of diffraction which may be vividly revealed by analysis of the intensity distribution in the diffracted field. The natural frequency of a resonator is given by $\omega_0 = (c^2 S/Vl')^{1/2}$, where V is the volume, S is the cross-sectional area of the neck and l' is the effective length of the neck. The other important parameter is the dynamic magnification, or Q-factor. Where energy dissipation is dominated by viscous flow phenomena, $Q = 2\omega_0 \rho_0 l'/R$, where R is the resistive force on the air in the neck produced by unit volume velocity of air through the neck; the equivalent internal damping ratio equals $1/2Q$.

When the wavelength of an incident sound field greatly exceeds the cross-sectional dimensions of the neck, the response of a resonator depends only on the incident sound pressure at the location of its mouth, and not on the spatial form of that field; in other words, it is a 'locally reacting' system. The response of a resonator to an incident field, and the consequent effect on that field, is critically dependent on the degree of matching between its internal resistance and its radiation resistance: in physical terms these may be defined, respectively, as the power dissipated within the resonator, and the power radiated into the surrounding fluid, per unit of volume velocity through the neck. The

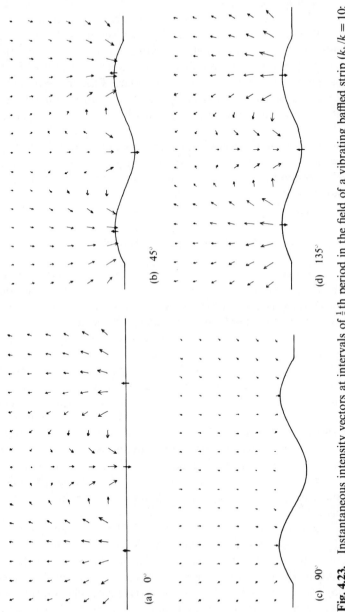

(a) 0°

(b) 45°

(c) 90°

(d) 135°

Fig. 4.23. Instantaneous intensity vectors at intervals of $\frac{1}{8}$th period in the field of a vibrating baffled strip ($k_b/k = 10$; vector scale $\propto I(t)^{1/4}$).

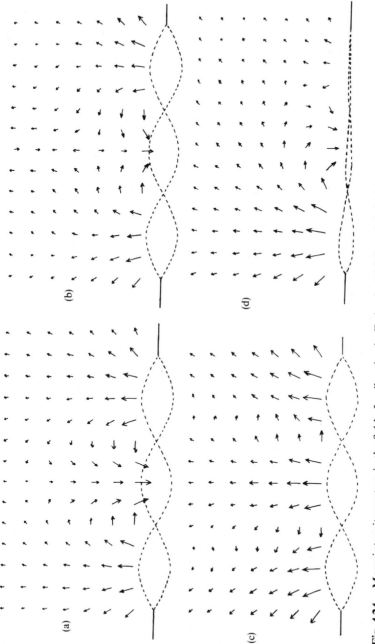

Fig. 4.24. Mean intensity vectors in the field of a vibrating baffled strip: (a) $k/k_b = 0.1$; (b) $k/k_b = 0.6$; (c) $k/k_b = 1.0$; (d) $k/k_b = 0.1$.

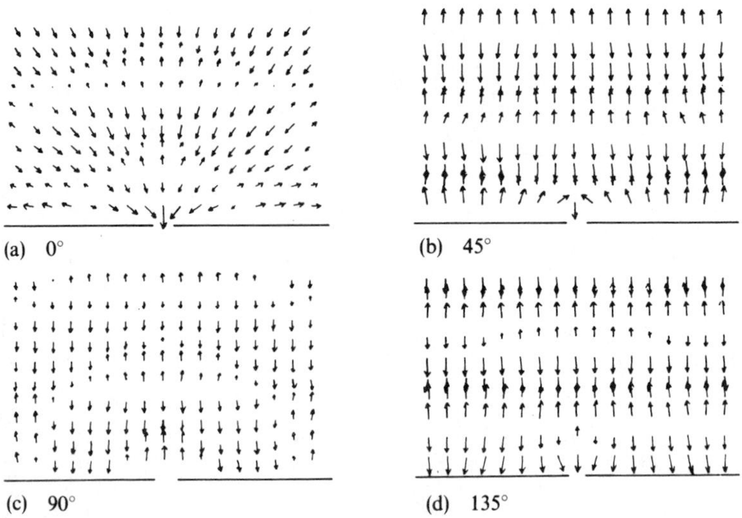

Fig. 4.25. Instantaneous intensity vector at intervals of $\frac{1}{8}$th period in the field of a plane wave normally incident upon a Helmholtz resonator at the resonance frequency ($r_{rad}/r = 0.4$).

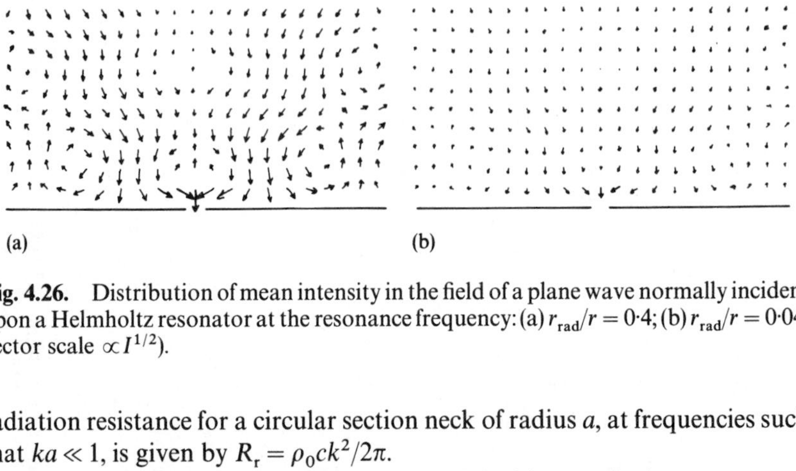

Fig. 4.26. Distribution of mean intensity in the field of a plane wave normally incident upon a Helmholtz resonator at the resonance frequency: (a) $r_{rad}/r = 0.4$; (b) $r_{rad}/r = 0.04$; vector scale $\propto I^{1/2}$.

radiation resistance for a circular section neck of radius a, at frequencies such that $ka \ll 1$, is given by $R_r = \rho_0 ck^2/2\pi$.

Intensity fields produced by the normal incidence of plane waves on a resonator at the resonance frequency are illustrated in Figs 4.25 and 4.26. The importance of matching the internal and radiation resistances for efficient absorption performance is clearly demonstrated by the difference between the magnitudes of the vectors 'entering' the neck (the lengths of the plotted vectors are proportional to $I^{1/2}$). An apparently paradoxical feature of the resonator is that it absorbs most strongly when it scatters the incident field most strongly.

The reactive component of intensity is not plotted because it is dominated by the interference between the incident wave and that reflected from the rigid baffle surrounding the resonator neck; however, detailed examination of the field in the vicinity of the neck reveals a strong convergence of reactive vectors, corresponding to low pressure at the mouth. The kinetic energy density is also very high in comparison with the potential energy density. The location of a resonator can most easily be detected by measuring the ratio of particle velocity to pressure with an intensity probe, which is far greater than $1/\rho_0 c$ in this region.

4.10   POWER FLUX STREAMLINES

The distribution of mean intensity (sound power flux density) in a steady, two-dimensional sound field may be represented by power flux streamlines, which are analogous to the streamlines of steady fluid flow. A streamline is defined to be any continuous line through the field across which there is zero mean power flow. It follows that the intensity vector at any point is tangential to the streamline which passes through that point. The field wavefronts are orthogonal to the streamlines.[37] Gauss's theorem, which leads to the conclusion that the divergence of the mean intensity vector is everywhere zero in the absence of sources or sinks, implies that the power flux through the channel formed between any two streamlines is conserved. Plots of streamlines

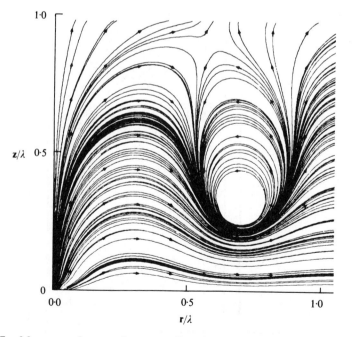

Fig. 4.27. Mean sound power flux streamlines in the field of a point-force excited, water-loaded plate.[46]

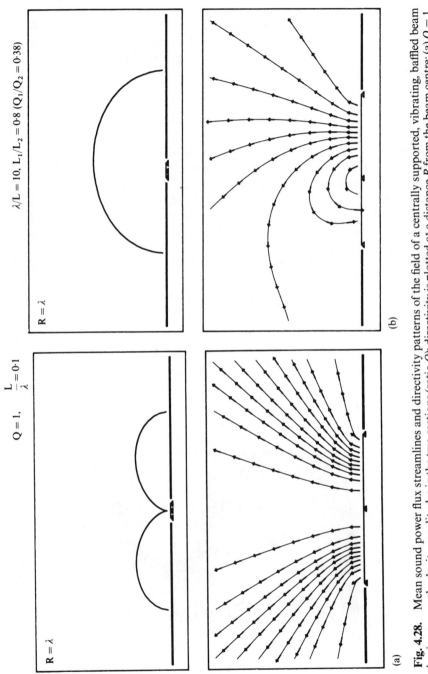

Fig. 4.28. Mean sound power flux streamlines and directivity patterns of the field of a centrally supported, vibrating, baffled beam having unequal velocity amplitudes in the two sections (ratio Q): directivity is plotted at a distance R from the beam centre: (a) $Q = 1$, $\lambda/L = 10$, $R = \lambda$; (b) $Q = 0.38$, $\lambda/L = 10$, $R = \lambda$; (c) $Q = 0.38$, $\lambda/L = 2$, $R = 5\lambda$; (d) $Q = 0.38$, $\lambda/L = 0.5$, $R = 20\lambda$.[47]

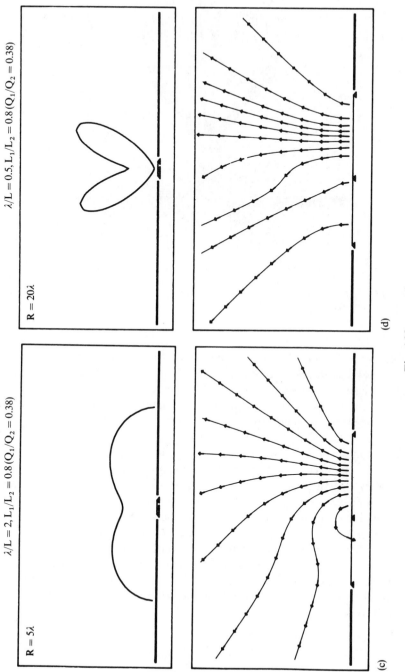

$\lambda/L = 0.5, L_1/L_2 = 0.8\,(Q_1/Q_2 = 0.38)$

$R = 20\lambda$

$\lambda/L = 2, L_1/L_2 = 0.8\,(Q_1/Q_2 = 0.38)$

$R = 5\lambda$

(d)

(c)

Fig. 4.28—*contd.*

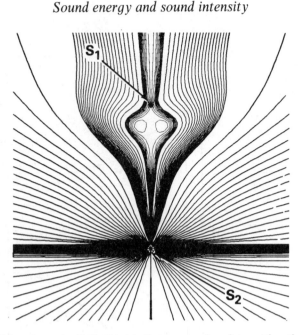

Fig. 4.29. Mean intensity field of two point monopoles of strength ratio 5 separated by a non-dimensional distance $kd = 5$.[48]

selected in such a way as to correspond to uniform increments of power flux (as with equal increments of the stream function in potential fluid flow) provide a clear visual display of the paths and concentrations of mean power flow. However, as demonstrated by Mann *et al.*,[37] such representations do not fully reveal the *process* by which the energy flows within a field. Some examples of theoretical streamline plots are presented in Figs 4.27–4.29.[46–48]

The example of the point-driven, water-loaded plate structure is especially interesting because it demonstrates the phenomenon of power flow into a vibrating structural sink (see also Fig. 10.15). A vibrating structure in contact with a fluid acts as a waveguide, redistributing such absorbed energy by transmission into connected structures, dissipation into heat, or re-radiation from other areas.

Streamline representation appears to be only appropriate to two-dimensional fields. Extension of the concept to general, three-dimensional fields has not, to the author's knowledge, been attempted. In fact, clear visualisation and graphical representation of three-dimensional intensity fields are extremely difficult to achieve.

CHAPTER FIVE

Principles of measurement of sound intensity

5.1 INTRODUCTION

In accordance with the definition of instantaneous sound intensity as the product of the instantaneous acoustic pressure and the instantaneous particle velocity, an intensity measurement system should, in principle, incorporate transducers of each of these two quantities. However, it transpires that direct transduction of particle velocity is not necessary, and one of the principal current commercial measurement systems implements an indirect transduction principle. It is, of course, imperative that the presence of the transducers distorts (diffracts) the sound field to an acceptably small degree, and that the transducer assembly does not vibrate at audio-frequencies with a velocity amplitude comparable with the particle velocity of the acoustic field (typically $10^{-4}\,\mathrm{m\,s^{-1}}$).

Equations (4.36a) and (4.36b) indicate that the magnitudes of the active and reactive components of intensity in a harmonic field may be obtained from the spatial gradients of phase and mean square pressure. These, together with corresponding relationships for acoustic impedance, were the subject of publications in the 1960s by Mechel,[11] Odin[12] and Kurze.[13] They are today effectively implemented in spectral form by FFT processing of signals from two closely spaced pressure microphones.

In this chapter, the principles of processing of the two signals from intensity probes will be presented, together with an analysis of the systematic errors which are inherent in the measurement principles employed. Systematic errors associated with non-ideal transducers and instrumentation performance are treated in Chapter 6, which deals with hardware and signal processing procedures, and in Chapter 7.

5.2 PRINCIPLES OF MEASUREMENT OF SOUND INTENSITY

Sound pressure and particle velocity in a sound field can both be expressed as functions of the velocity potential of the field, but the relationship between the two depends upon the type of sound field, and is not unique. Therefore it is necessary to employ at least two transducers to determine sound intensity. Pairs of transducers are physically associated in an intensity 'probe'. Two categories of probe are currently in use; one combines a pressure transducer with a particle velocity transduction unit; the other comprises two, nominally identical, pressure transducers (microphones). We shall henceforth refer to the former as a 'p-u' probe, and to the latter as a 'p-p' probe, which enjoys far greater use.

5.2.1   The p-u principle

The two output signals from a probe which incorporates a combination of a pressure transducer and a particle velocity transducer are multiplied together to give the time-dependent component of intensity in the direction of the probe axis. The only current, commercially available p-u probe combines a standard condenser microphone with an ultrasonic particle velocity transducer, as shown in Fig. 6.8 in Chapter 6. Two parallel ultrasonic beams are launched in opposing directions; convection of these waves by the oscillatory air movement of any audio-frequency wave present in the same fluid produces a difference of phase between the two ultrasonic waves on arrival at their respective receptors. This phase difference is an analogue of the particle velocity component of the audio-frequency wave in the direction of the beams. Of course, this system also responds to any non-acoustic air movements, such as those of turbulence in wind, and therefore precautions must be taken to screen the probe from such disturbances.

In the absence of superimposed airflow, the transit time of each beam is the same, and is given by $t_0 = d/c$, where d is the distance between the faces of the sender and receptor. If a steady flow of speed u is superimposed, the transit times become $t^+ = d/(c + u)$ and $t^- = d/(c - u)$. The resulting phase difference is

$$\delta\phi = \omega_u d[1/(c - u) - 1/(c + u)] \approx 2\omega_u \, du/c^2 \quad \text{if } u \ll c$$

ω_u is the ultrasonic frequency. This phase difference is converted into an electrical analogue of u. If the superimposed flow is not uniform in space, but due to the presence of an audio-frequency sound wave, the relationships between the transit times and the particle velocity component at the central microphone position take complicated forms, which may be evaluated numerically: the details are presented in Section 5.6.2.

The signal proportional to pressure may be directly multiplied by the signal proportional to the particle velocity component to produce an analogue of the instantaneous intensity component. If appropriate, this intensity signal can be time-averaged to indicate the mean (active) intensity component. Frequency decomposition of the intensity may be achieved either by identically filtering the two signals before multiplication, or, as we shall see in Section 5.3, spectral analysis may be employed. The total instantaneous intensity vector $\mathbf{I}(t)$ can only be determined if three orthogonal components of particle velocity can be simultaneously measured. However, if only the mean intensity vector in a steady sound field is required, the results of sequential measurements of the three mean intensity components may be vectorially combined.

5.2.2 The p-p principle

Two nominally identical pressure transducers are placed close together in a support structure which is designed to minimise diffraction of the incident sound field. The transducers are normally high quality condenser microphones for measurements in air, and piezo-electric hydrophones for measurements in water. Most condenser microphone capsules take the form of short cylinders which may be associated in various configurations, including 'face-to-face', 'side-by-side', 'tandem' and 'back-to-back' (Fig. 5.1). A signal proportional to the component of particle velocity which is co-linear with the probe axis (i.e. the line joining the acoustic centres of the transducers) is obtained by employing a finite difference approximation to the local spatial gradient of sound pressure. The zero mean flow fluid momentum equation (3.12) shows that, in a small amplitude sound field, the component of pressure gradient in any direction n is proportional to the component of fluid particle acceleration in that direction:

$$\partial p/\partial n = -\rho_0 \partial u_n/\partial t \tag{5.1}$$

The corresponding component of particle velocity is therefore given by the time integral

$$u_n(t) = -(1/\rho_0) \int_{-\infty}^{t} (\partial p(\tau)/\partial n)\, d\tau \tag{5.2}$$

This is approximated as

$$u_n(t) \approx (1/\rho_0 d) \int_{-\infty}^{t} [p_1(\tau) - p_2(\tau)]\, d\tau \tag{5.3}$$

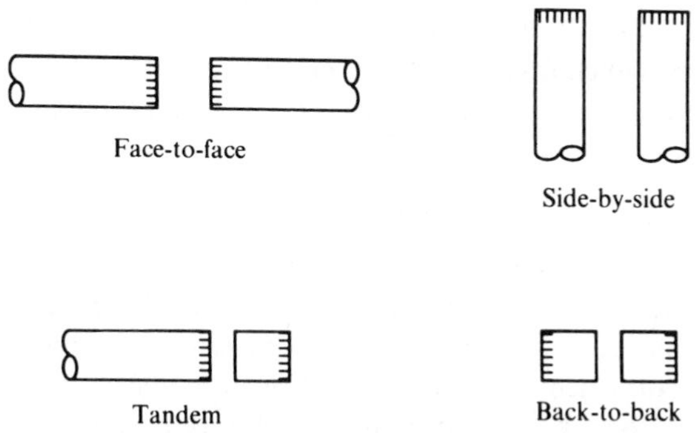

Face-to-face

Side-by-side

Tandem Back-to-back

Fig. 5.1. Schematic p-p intensity probe configurations.

where d is the distance separating the acoustic centres of the transducers; this will henceforth be termed the 'separation distance'.

The pressure at the point midway between the transducers is approximated as

$$p(t) \approx 1/2[p_1(t) + p_2(t)] \tag{5.4}$$

Hence, the instantaneous intensity component is approximated by

$$I_n(t) \approx (1/2\rho_0 d)[p_1(t) + p_2(t)] \int_{-\infty}^{t} [p_1(\tau) - p_2(\tau)] \, d\tau \tag{5.5}$$

The remarks in the previous section concerning the evaluation of the total instantaneous and mean intensity vectors apply equally to p-p measurements.

Equation (5.5) may be variously implemented using sum, difference, integration and multiplication circuits. In cases of temporally non-stationary sound fields such as those generated by transient sources, the evolution of $I(t)$ is of vital interest, and the full equation must be implemented. However, many sources operate steadily; their sound fields may be considered to be stationary, and for the determination of source sound power the mean intensity is of prime interest. Time stationary signals $x(t)$ and $y(t)$ are such that $\overline{x(dx/dt)} = \overline{y(dy/dt)} = 0$, and $\overline{x(dy/dt)} = -\overline{y(dx/dt)}$. In this case, p_1 is the time derivative of $\int p_1 \, d\tau$, and the mean (active) intensity component in direction n is given by

$$I_n = -(1/\rho_0 d) \lim_{T \to \infty} (1/T) \int_0^T \left[p_1(t) \int_{-\infty}^{t} p_2(\tau) d\tau \right] dt \tag{5.6}$$

Hence, in principle, the mean intensity in a time-stationary sound field can be obtained from the product of the signal from one pressure transducer and the integrated signal from another identical transducer in close proximity. For example, in the case of simple harmonic signals $p_1 = P_1 \exp[i(\omega t + \phi_1)]$ and $p_2 = P_2 \exp[i(\omega t + \phi_2)]$, the intensity is given by

$$I_n = P_1 P_2 \sin(\phi_1 - \phi_2)/2\rho_0 \omega d \approx P_1 P_2 (\phi_1 - \phi_2)/2\rho_0 \omega d \quad \phi_1 - \phi_2 \ll 1 \tag{5.7}$$

Measurements could therefore be made with only one microphone, which is located in turn at two suitably spaced points, together with a high resolution phase meter, by which each signal is referred to a stable reference signal.

The magnitude of the associated oscillatory, reactive flow of sound energy in the direction n is, from eqn (4.29),

$$|J| = |I_{re}| \approx [P_1^2 - P_2^2]/4\omega\rho_0 d \tag{5.8}$$

For accurate implementation of the expressions in eqns (5.5)–(5.8), it is obvious that the two pressure transducers must have identical impulse responses within the frequency range of measurement, and any integrator must also perform accurate a.c. integration. For measurements in steady sound fields, d.c. bias, but not drift, in the integration circuit can be tolerated, because its output is multiplied by a signal having zero mean value.

5.3   FREQUENCY DISTRIBUTION OF SOUND INTENSITY
IN TIME-STATIONARY SOUND FIELDS

As we have seen, the component of the instantaneous sound intensity in any particular direction is a time-dependent quantity. The expression relevant to harmonic fields is eqn (4.27). Clearly, if such harmonic fluctuations were subjected to Fourier analysis, the active intensity component would yield a d.c. component of magnitude $|I_a|$, plus a harmonic component of amplitude $|I_a|$ at frequency 2ω. The reactive intensity would also yield a harmonic component of amplitude $|J|$ at frequency 2ω, in quadrature with the active component. The flow of sound energy is actually produced by 'co-operation' between harmonic components of pressure and particle velocity, each having the same frequency ω; but the Fourier analysis would only indicate activity at frequencies 2ω and zero. Superimposition of another harmonic field of a different frequency would appear to confuse the situation further; there would be fluctuations of intensity at sum and difference frequencies, as well as at twice the harmonic frequencies.

These considerations serve to show that:

(i)   if we wish to apportion mean intensity in a steady field according to a distribution in frequency, it is the magnitudes and relative phase of components of pressure and particle velocity of the *same* frequency which determine the apportionment to that frequency;

(ii) the frequency distribution of time-varying intensity in a steady, multi-frequency field is of no apparent practical significance.

By definition, the component of the mean intensity vector in any direction is equal to the time-average product of the pressure and the component of the particle velocity in that direction. Hence the distribution of mean intensity according to frequency may be obtained directly from the output of a p-u probe by passing the two signals through identical filters prior to forming the time-average product. In the case of a p-p probe, the frequency distribution may be determined according to the time-average form of eqn (5.5) by passing the two signals through identical filters, either before or after performing the sum, difference and integration operations, and then performing the time-average operation on the product of the filtered outputs. The choice of the stage at which filtering is applied is largely dictated by the need to optimise hardware performance. In the case of stationary signals, eqn (5.6) may be similarly implemented, thereby avoiding sum and difference operations. Application of these procedures may be termed 'direct' frequency analysis.

'Indirect' frequency analysis procedures are based upon Fourier (spectral) analysis of the two probe signals, which is here introduced via the correlation function which indicates the time-average relationship between two signals in the time domain. The Cross-correlation Function between the pressure and particle velocity is defined as

$$R_{pu}(\tau) = \lim_{T \to \infty} (1/T) \int_0^T p(t)u(t + \tau) \, dt$$

(Note: in the case of harmonic signals the limiting process is replaced by a time average over an integer number of cycles.)

Hence, the mean intensity component in direction **r** is given by

$$I_r = \lim_{T \to \infty} (1/T) \int_0^T p(t)u_r(t) \, dt = R_{pu}(0) \qquad (5.9)$$

in which the directional subscript on u in the last term has been dropped for typographical clarity. The distribution in frequency of the product of the p and u components of that frequency is given by the Fourier transform of the cross-correlation function, which is termed the Cross-spectral Density:

$$S_{pu}(\omega) = \frac{1}{2\pi} \int_{-\infty}^{\infty} R_{pu}(\tau) \exp(-i\omega\tau) \, d\tau \qquad (5.10)$$

This function is mathematically complex, indicating the average phase relationship between p and u.

$R(\tau)$ and $S(\omega)$ form a Fourier transform pair, and thus

$$R_{pu}(\tau) = \int_{-\infty}^{\infty} S_{pu}(\omega) \exp(i\omega\tau) \, d\omega$$

and

$$I_r = R_{pu}(0) = \int_{-\infty}^{\infty} S_{pu}(\omega) \, d\omega \qquad (5.11)$$

In this sense, $S_{pu}(\omega)$ represents the distribution of the contributions of different frequency components of the sound field to the mean intensity. Cross-spectra possess the following properties:

$$\mathrm{Re}\{S_{pu}(\omega)\} = \mathrm{Re}\{S_{pu}(-\omega)\} \qquad (5.12a)$$
$$\mathrm{Im}\{S_{pu}(\omega)\} = -\mathrm{Im}\{S_{pu}(-\omega)\} \qquad (5.12b)$$

The spectral function $S(\omega)$ is defined for all positive and negative frequencies, i.e. it can be represented by pairs of counter-rotating phasors. For practical purposes it is convenient to redefine the spectral densities as single-sided functions of frequency, thus:

$$G_{pu}(\omega) = 2S_{pu}(\omega) \quad \omega > 0$$
$$G_{pu}(\omega) = S_{pu}(\omega) \quad \omega = 0$$
$$G_{pu}(\omega) = 0 \quad \omega < 0$$

Hence, the distribution of contributions of the different frequency components to the mean intensity component is

$$I_r(\omega) = S_{pu}(\omega) + S_{pu}(-\omega)$$
$$= 2 \operatorname{Re} \{S_{pu}(\omega)\}$$
$$= \operatorname{Re} \{G_{pu}(\omega)\} \tag{5.13}$$

When a p-u probe is used, eqn (5.13) may be implemented directly with a two-channel FFT analyser to give $I_r(\omega)$ in the direction of the probe axis. The total vector in a stationary field may be obtained by vector addition of the results of sequential measurements in three orthogonal directions.

The imaginary part of $G_{pu}(\omega)$ is proportional to the magnitude of the reactive intensity: however, unlike the real part, it does not represent the distribution of contributions of frequency components to a time-average quantity because the mean reactive intensity is zero at all frequencies (eqn (5.12b)).

The spectral analysis of signals from a p-p probe is more involved. The intensity component in the direction of the probe axis is given by eqn (5.6)

$$I_r = -(1/\rho_0 d) \lim_{T\to\infty} (1/T) \int_0^T p_1(t) \left[\int_{-\infty}^t p_2(\tau') \, d(\tau') \right] dt$$

in which τ has been replaced by τ' to avoid confusion with the correlation time delay. If we write $\int p_2(\tau') \, d\tau'$ as $z_2(t)$, the appropriate cross-correlation function is

$$R_{pz}(\tau) = \lim_{T\to\infty} (1/T) \int_{-\infty}^T p_1(t) z_2(t + \tau) \, dt \tag{5.14}$$

and the corresponding cross-spectral density function is

$$S_{pz}(\omega) = \frac{1}{2\pi} \int_0^\infty R_{pz}(\tau) \exp(-i\omega\tau) \, d\tau \tag{5.15}$$

By analogy with eqn (5.13),

$$I_r(\omega) = -(2/\rho_0 d) \operatorname{Re} \{S_{pz}(\omega)\} \tag{5.16}$$

Rewriting $\exp(-i\omega\tau)$ as $\exp(-i\omega(t + \tau) + i\omega t)$, and $d\tau$ as $d(t + \tau)$ for fixed t, eqn (5.15) becomes

$$S_{pz}(\omega) = \frac{1}{2\pi} \int_{-\infty}^\infty \left[\lim_{T\to\infty} (1/T) \int_0^T p_1(t) z_2(t + \tau) \, dt \right]$$
$$\times \exp(-i\omega(t + \tau)) \exp(i\omega t) \, d(t + \tau) \tag{5.17}$$

Considering first the integral with respect to $t + \tau$ at fixed t:

$$\int_{-\infty}^\infty z_2(t + \tau) \exp(-i\omega(t + \tau)) \, d(t + \tau)$$

$$= (i/\omega) z_2(t + \tau) \exp(-i\omega(t + \tau))|_{-\infty}^\infty - (i/\omega)$$

$$\times \int_{-\infty}^\infty (dz_2/d(t + \tau)) \exp(-i\omega(t + \tau)) \, d(t + \tau)$$

of which the first term is zero. Now, $dz_2/d(t + \tau) = p_2(t + \tau)$. Hence, eqn (5.17) may be written

$$S_{pz}(\omega) = - (i/2\pi\omega) \int_{-\infty}^{\infty} \left[\lim_{T \to \infty} (1/T) \int_0^T p_1(t)p_2(t + \tau)\,dt \right] \exp(-i\omega\tau)\,d\tau$$

$$= - (i/2\pi\omega) \int_{-\infty}^{\infty} R_{p1p2}(\tau)\exp(-i\omega\tau)\,d\tau$$

$$= - (i/\omega)S_{p1p2}(\omega) \tag{5.18}$$

Therefore

$$\mathrm{Re}\,\{S_{pz}(\omega)\} = (1/\omega)\,\mathrm{Im}\,\{S_{p1p2}(\omega)\}$$

and

$$I_r(\omega) = - (2/\rho_0\omega d)\,\mathrm{Im}\,\{S_{p1p2}(\omega)\}$$

or

$$I_r(\omega) = (1/\rho_0\omega d)\,\mathrm{Im}\,\{G_{p2p1}(\omega)\} \tag{5.19}$$

which is simply the spectral form of eqn (5.6).

The great practical utility of this expression is that it can be implemented simply by feeding the outputs of two well-matched microphones directly into an FFT analyser.

Equation (5.19) may also be derived directly in the frequency domain by employing the Fourier transforms of p_1 and $\int p_2$, denoted by $P_1(\omega)$ and $Z_2(\omega)$. The cross-spectral density $G_{pu}(\omega)$ is a limited function of the (unlimited) Fourier transforms of pressure $P(\omega)$ and particle velocity $U(\omega)$;

$$G_{pu}(\omega) = \lim_{T \to \infty} (2/T)[P^*(\omega)U(\omega)] \tag{5.20}$$

Fourier transformation of the fluid momentum equation (3.12) yields

$$F\{\partial p/\partial r\} = - i\omega\rho_0 U_r(\omega)$$

Now, from the finite difference approximation (eqn (5.3))

$$[P_1(\omega) - P_2(\omega)]/d \approx i\omega\rho_0 U_r(\omega)$$

and from eqn (5.4)

$$P(\omega) \approx \tfrac{1}{2}[P_1(\omega) + P_2(\omega)]$$

Hence

$$G_{pu}(\omega) = - (i/2\rho_0\omega d) \lim_{T \to \infty} (2/T)\{[P_1^*(\omega) + P_2^*(\omega)][P_1(\omega) - P_2(\omega)]\}$$

$$= - (i/2\rho_0\omega d) \lim_{T \to \infty} (2/T)\{|P_1(\omega)|^2 - |P_2(\omega)|^2$$

$$+ P_2^*(\omega)P_1(\omega) - P_1^*(\omega)P_2(\omega)\}$$

$$= - (i/2\rho_0\omega d)[G_{p1p1}(\omega) - G_{p2p2}(\omega) - G_{p1p2}(\omega)$$

$$+ G_{p2p1}(\omega)] \tag{5.21}$$

Equation (5.13) gives the spectral density of the mean intensity as the real part of G_{pu}: thus

$$I_r(\omega) = -(1/\rho_0 \omega d)\, \mathrm{Im}\,\{G_{p1p2}(\omega)\} \qquad (5.22)$$

in agreement with eqn (5.19).

The imaginary part of G_{pu} indicates the amplitude of the reactive intensity:

$$J_r(\omega) = -\,\mathrm{Im}\,\{G_{pu}(\omega)\} = (1/2\rho_0 \omega d)[G_{p1p1}(\omega) - G_{p2p2}(\omega)] \qquad (5.23)$$

in which the minus sign before $\mathrm{Im}\,\{G\}$ is necessary for compatibility with the definition of J for harmonic fields in eqn (4.34d). Since G_{pp} is the auto(power) spectral density of squared pressure, this result agrees with the harmonic form (eqn (5.8)), in which $\overline{p^2} = P^2/2$. The sign is of practical significance since it may be used to distinguish the diverging patterns of reactive intensity in the near fields of sources from the converging patterns observed near pressure minima in interference fields, and near other regions of low acoustic pressure.

The amplitude of the reactive intensity may be determined by implementing eqn (5.23) with an FFT analyser. Alternatively, a simple sound level meter may be used to determine the differences of mean square pressures in 1/1 or 1/3 octave frequency bands.

5.4   MEASUREMENT OF SOUND ENERGY FLUX OF TRANSIENT SOUND FIELDS

The total sound energy flux per unit area having unit vector **n** during a time interval t_1 to t_2 is given by[49]

$$E = \int_{t_1}^{t_2} \mathbf{I.n}\, dt \qquad (5.24)$$

Equation (5.5) expresses the instantaneous intensity as

$$I_n(t) = (1/2\rho_0 d)(p_1(t) + p_2(t)) \int_{-\infty}^{t} (p_1(\tau) - p_2(\tau))\, d\tau \qquad (5.25)$$

The energy flux which occurs in the time interval up to time T is given by substituting the expression for intensity given by eqn (5.25) into eqn (5.24) as

$$E(T) = (1/2\rho_0 d) \int_{-\infty}^{T} \left[(p_1(t) + p_2(t)) \int_{-\infty}^{t} (p_1(\tau) - p_2(\tau))\, d\tau \right] dt \qquad (5.26)$$

Consider the integral

$$\int_{-\infty}^{T} p(t) \int_{-\infty}^{t} p(\tau)\, d\tau\, dt$$

It can be written as

$$\int_{-\infty}^{T} X(dX/dt)\, dt = [X^2/2]_{-\infty}^{T} = 1/2[X^2(T) - X^2(-\infty)]$$

where $X(t) = \int_{-\infty}^{t} p(\tau) \, d\tau$. Now, a transient event by definition occupies a finite interval of time, and for all normal, small amplitude acoustic disturbances in otherwise quiescent air

$$\int_{-\infty}^{T} p(t) \, dt = 0$$

provided that the fluid has returned to its equilibrium condition and position by time T. Hence eqn (5.26) for the total energy flux of the event may be simplified to

$$E = -(1/2\rho_0 d) \int_{T_1}^{T_2} \left[p_1(t) \int_{-\infty}^{t} p_2(\tau) \, d\tau - p_2(t) \int_{-\infty}^{t} p_1(\tau) \, d\tau \right] dt \quad (5.27)$$

in which the transient disturbance concerned starts after time T_1 and ceases to exist before time T_2. This condition refers to the sound field at the measurement point, not to the source operation. The condition

$$\int_{-\infty}^{T_2} p(t) \, dt = 0$$

allows eqn (5.27) to be simplified further to

$$E = (1/\rho_0 d) \int_{T_1}^{T_2} \left[p_2(t) \int_{-\infty}^{t} p_1(\tau) \, d\tau \right] dt \quad (5.28)$$

Equation (5.28) is only valid as an estimate of the *total* energy flux of the event: it is not valid for any portion of an event within time limits at which p_1 and p_2 are non-zero.

By representing $p_1(t)$ and $p_2(t)$ in eqn (5.28) by their finite Fourier transforms $P_1(f)$ and $P_2(f)$, the spectral distribution of the total energy flux may be shown to be given by

$$E(f) = (2/\rho_0 c k d) \, \mathrm{Im} \, \{ P_1(f) P_2(f)^* \} \quad (5.29)$$

where

$$E = \int_0^{\infty} E(f) \, df$$

This finite transform equivalent of the cross-spectral eqn (5.19) is only valid when applied to the whole transient event; it may not be used to evaluate the temporal evolution of the energy flux spectrum.

The influence of extraneous noise during transient intensity measurement presents a more serious problem than it does for measurements in stationary fields, essentially because extraneous disturbances and their associated energy fluxes will not normally satisfy the condition of finite duration on which eqn (5.28) is based. The contributions to the energy flux of 'co-operation' between source field pressures and extraneous field particle velocities, and vice versa, might reasonably be expected to be small, provided that the transient event is not extremely brief: but the contribution of a continuous extraneous field is not likely to integrate to zero over the duration of a transient event. A means of suppressing such contributions is suggested in Ref. 49: finite Fourier transforms are evaluated over two different time intervals, one being twice the other and both encompassing the transient event. The transforms of continuous stationary noise will double, whereas those of the transient signals will not change. The resulting expression for $E(f)$ is

$$E(f) = 2E'(f)_2 - E'(f)_1 \qquad (5.30)$$

in which $E'(f)_i$ are the biased estimates.

5.5   SURFACE SOUND INTENSITY MEASUREMENT

In parallel with the development of routine sound intensity measurement by means of p-p and p-u probes, a number of investigations were made of the radiation characteristics of vibrating structures by the use of combinations of pressure microphones and surface vibration transducers.[50-53] Vibration is most commonly measured with accelerometers, but optical probes have also been employed. The microphone is placed in close proximity to the point of vibration measurement, and the local normal intensity is determined by using the various p-u signal processing procedures described in Section 5.3.

The technique is of particular value under conditions of extremely high extraneous noise which seriously degrade the accuracy of airborne intensity measurements; however, it has some disadvantages: (i) traversing a combination of surface velocity and pressure transducers is awkward; (ii) near field measurements are less than ideal for the determination of sound power radiated by vibrating sources; (iii) vibration fields are less uniform, and therefore more difficult to sample, than sound fields; (iv) leaks in partitions cannot be detected.

5.6   SYSTEMATIC ERRORS INHERENT IN SOUND INTENSITY MEASUREMENT TECHNIQUES

The two different techniques currently used for the transduction of sound pressure and particle velocity are subject to systematic errors which arise from the fact that they involve approximations which are inherent in the transduction principles employed. The errors inherent to the p-p technique are different in nature from those inherent in the p-u technique: the former arise from the approximations expressed in eqns (5.3) and (5.4); the latter arise specifically from the operating principle of the ultrasonic particle velocity transducer system. The resulting errors are analysed here, rather than in the following chapters on instrumentation and measurement performance, because they result directly from the principles of intensity measurement employed, and not from imperfections in the measurement systems *per se*. Inherent errors are,

however, functions of the type of field under investigation, and of the orientation of the probe within the field. The slightly disturbing implication of this fact
is that the magnitude of an inherent error can never be precisely estimated in
an arbitrary sound field, the properties of which are not known *a priori*.
Therefore examples of errors in a range of idealised sound fields are presented
in order to provide an indication of their sensitivity to the parameters of the
field and of the measurement probe.

5.6.1   Systematic errors inherent in the p-p technique

As mentioned above, the errors inherent in the application of the p-p technique
derive from the finite difference approximations to the pressure and axial
particle velocity component expressed in eqns (5.3) and (5.4). They are most
easily analysed by reference to the spatial distribution of the pressure field in
the direction of the probe axis, which for convenience will be denoted by the
co-ordinate x. According to the Taylor series expansion

$$p(x + h, t) = p(x, t) + hp'(x, t) + (h^2/2)p''(x, t) + (h^3/6)p'''(x, t)$$
$$+ \cdots + (h^n/!n)p^n(x, t) + \cdots \tag{5.31}$$

where $p(x, t)$ has arbitrary time dependence and $p^n(x, t)$ denotes the nth
derivative of p with respect to x at any instant t.

Let us consider a pair of pressure transducers of which the *acoustic centres*
are separated by a distance $2h$. (Note: the acoustic centre refers to that point at
which the field pressure most closely corresponds to the pressure transducer
output: it normally varies with frequency, and with the direction of the incident
sound wave; therefore the following analyses should strictly only be applied at
single frequencies, a fact which invalidates the direct application of the p-p
technique to the measurement of transient sound fields. However, the variations
of h may be made negligibly small by good probe design.) Equation (5.4) gives
the estimated pressure at the point midway between the transducer centres as

$$p_e(t) = p(t) + (h^2/2)p''(t) + (h^4/24)p^{iv}(t) + \cdots \tag{5.32}$$

in which explicit indication of spatial position has been dropped. From eqn
(5.3), the estimated axial particle velocity component at the centre of the probe
is given by

$$u_e(t) = -(1/\rho_0) \int_{-\infty}^t [p'(\tau) + (h^2/6)p'''(\tau) + (h^4/120)p^v(\tau) + \cdots] \, d\tau \tag{5.33}$$

Hence the normalised errors in the estimates of p and u are

$$e(p) = (p_e - p)/p = [(h^2/2)p''(t) + (h^4/24)p^{iv}(t) + \cdots]/p(t) \tag{5.34}$$

and

$$e(u) = (u_e - u)/u = \int_{-\infty}^t [(h^2/6)p'''(\tau) + (h^4/120)p^v(\tau) + \cdots] \, d\tau \bigg/ \int_{-\infty}^T p'(\tau) \, d\tau \tag{5.35}$$

These expressions cannot be evaluated unless the time history of p and of its spatial derivatives is known. Therefore we now specialise to harmonic fields of frequency ω, in which $p(x,t) = \text{Re}\{P(x)\exp(i\omega t)\}$ and $u(x,t) = \text{Re}\{U(x)\exp(i\omega t)\}$; again the explicit indication of x-dependence is omitted. Now

$$U = (i/\omega\rho_0)P'$$
$$U_e = -(i/2\omega\rho_0 h)(P_1 - P_2)$$

and

$$e(u) = [(h^2/6)P''' + (h^4/120)P^v + \cdots]/P'(t) \tag{5.36}$$

The estimated intensity is, from eqn (5.6), given by

$$I_e = (1/4\rho_0\omega h)\,\text{Im}\{P_1 P_2^*\} \tag{5.37}$$

The true intensity is given by

$$I = \tfrac{1}{2}\text{Re}\{PU^*\} = (1/2\omega\rho_0)\,\text{Im}\{PP'^*\} \tag{5.38}$$

Taylor series expansion of eqn (5.37) yields

$$I_e = (1/4\omega\rho_0 h)\,\text{Im}\{h[PP'^* - P^*P'] + (h^3/6)[PP'''^* - P^*P''']$$
$$- (h^3/2)[P'P''^* - P'^*P'''] + \cdots\} \tag{5.39}$$

The errors associated with some common idealised models of sound fields are now analysed, as first done by Fahy[20] and Pavic.[19]

(a) Plane progressive wave

$$P = A\exp(-ikx)$$
$$P' = -ikP$$
$$P'' = -k^2 P$$
$$P''' = -k^2 P' = ik^3 P, \text{ etc.}$$

$$e(p) = \cos(kh) - 1 \approx -(kh)^2/2 + (kh)^4/24 - (kh)^6/720 + \cdots \tag{5.40a}$$

$$e(u) = (\sin(kh)/kh) - 1 \approx -(kh)^2/6 + (kh)^4/120 - (kh)^6/5040 + \cdots \tag{5.40b}$$

Note that the normalised error in the estimate of p is greater than that of u, and that only amplitude error, and not phase error, is produced by the finite difference approximation. The consequent normalised error in the sound intensity estimate is

$$e(I) = (I_e - I)/I \approx -(2/3)(kh)^2 + (2/15)(kh)^4 \tag{5.40c}$$

where it has been assumed that $kh \ll 1$. The logarithmic forms of $e(p)$, $e(u)$ and $e(I)$ for air at 20 °C are plotted as functions of the product of frequency and transducer separation ($d = 2h$) in Fig. 5.2.

The conditions for the normalised error to be less than 5%, $(10\lg(1 + e) = -0.2\,\text{dB})$ in terms of the non-dimensional p-p transducer

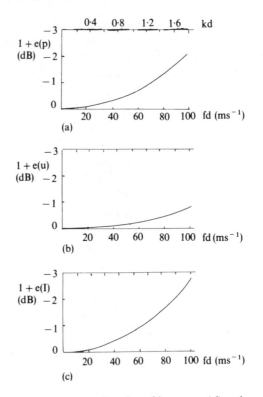

(a)

(b)

(c)

Fig. 5.2. Systematic p-p errors as a function of frequency (f) and separation distance (d): (a) pressure error; (b) particle velocity error; (c) intensity error.

separation distance $kd = 2kh$, are as follows:

$$e(p) < 5\% \quad kd < 0.66 \quad fd < 36 \tag{5.41a}$$
$$e(u) < 5\% \quad kd < 1.1 \quad fd < 60 \tag{5.41b}$$
$$e(I) < 5\% \quad kd < 0.55 \quad fd < 30 \tag{5.41c}$$

Note that the condition for $e(I)$ is twice as restrictive as for $e(u)$.

(b) Plane wave interference field

$$P(x) = A \exp(-ikx) + B \exp(ikx)$$

The ratios P''/P, P'''/P', etc., are all of the same form as for the progressive plane wave, and so eqns (5.40) and (5.41) apply.

(c) Point monopole field

$$P(r) = (A/r) \exp(-ikr)$$
$$P' = -[1/r + ik]P$$

$$P'' = [2/r^2 - k^2 + 2ik/r]P$$
$$P''' \approx [6/r^2 + k^2]P' \quad kr \ll 1$$
$$= -[6/r^3 - 3k^2/r + i(6k/r^2 - k^3)]P$$

The existence of an imaginary part in the ratio P''/P indicates that the finite difference approximation introduces both phase and amplitude errors into the estimate of P, but the ratio P'''/P' shows no phase error in the estimate of u. The normalised error in the estimate of p for small kr is

$$e(p) \approx -(kh)^2/2 + (h/r)^2 + i(kh)/(h/r) \quad kr \ll 1 \tag{5.42a}$$

The first term is the same as for the plane wave, as it must be for large r, but the other two involve the ratio of transducer separation distance to the distance between the centres of the source and the probe. The phase error depends upon both h/r and kr.

The normalised error in the estimate of u for small kr is

$$e(u) \approx (kh)^2/6 + (h/r)^2 \quad kr \ll 1 \tag{5.42b}$$

The first term is different from that for the plane wave; it is smaller than the corresponding term in $e(p)$, but the source proximity terms are the same: there is no significant phase error.

The normalised error in the estimate of I for small kr is

$$e(I) \approx -2(kh)^2/3 + (h/r)^2 \quad kr \ll 1 \tag{5.42c}$$

The first term is again the same as for the plane wave. The primary source proximity term contributes less than $2 \cdot 5\%$ error if $d/r < 0 \cdot 22$, i.e. with the centre of the probe not closer to the source centre than four times the transducer separation distance. This restriction would have practical significance if, for example, a large transducer separation were being used in the low frequency field of a bass loudspeaker unit.

In the far field, where $kr \gg 1$, the errors revert to the plane wave values.

(d) Point dipole field

$$p = M[ik/r + 1/r^2]\cos\theta \exp[i(\omega t - kr)] = P(r, \theta)\exp(i\omega t)$$

where M is the dipole moment source strength. Consider the radial component of pressure and particle velocity (the prime denotes differentiation with respect to r). For $kr \ll 1$,

$$P' \approx -[2/r + ik]P$$
$$P'' \approx [6/r^2 - k^2 + 2ik/r]P$$
$$P''' \approx [12/r^2 - k^2 - 5ik/r]P'$$

Hence

$$e(p) \approx -(kh)^2/2 + 3(h/r)^2 + i(kh)(h/r) \tag{5.43a}$$

$$e(u) \approx -(kh)^2/6 + 2(h/r)^2 - 5i(kh)(h/r)/6 \tag{5.43b}$$

and

$$e(I) \approx - 2(kh)^2/3 + 7(h/r)^2/3 \qquad (5.43c)$$

The source proximity error which depends on (h/r) is considerably larger for the dipole than for the monopole.

As with the monopole, the errors take the plane wave values in the far field, where the approximations based on an assumption of $kr \ll 1$ are not valid.

(e)   Flexural wave near field

The sound field generated by a uniform plate vibrating harmonically in flexure at frequencies well below its critical frequency may be represented approximately by that of an infinite plate field, expressed by eqn (4.73b) as

$$P(x, y) = (i\rho_0 \omega V_n/k_y) \exp(- k_y y) \exp(- ik_t x)$$

in which k_t is the wavenumber of the flexural wave, V_n is its normal velocity amplitude and $k_y = (k_t^2 - k^2)^{1/2}$. The normal components of the spatial derivatives of pressure take the same forms as for plane waves, but with k replaced by k_y:

$$P'' = - k_y^2 P \quad P''' = - k_y^2 P' \quad P^{iv} = k_y^4 P \quad \text{and} \quad P^v = k_y^4 P'$$

Hence

$$e(p) \approx - (k_y h)^2/2 + (k_y h)^4/24 \qquad (5.44a)$$

and

$$e(u_n) \approx - (k_y h)^2/6 + (k_y h)^4/120 \qquad (5.44b)$$

Figure 5.2 applies with kd replaced by $k_y d$.

Note that $e(p)$ is approximately three times $e(u_n)$. The normal component of mean intensity is zero, and therefore a normalised error estimate would be meaningless. In fact, the actual error due to finite separation is zero since the two sensed pressures are in phase. When $k_t \gg k$, $k_y \approx k_t$, and therefore the above errors are functions of the transducer separation normalised on the plate wavenumber and not on the acoustic wavenumber. Hence they may substantially exceed the plane wave errors at the same frequency.

There is always some radiation of power from vibrating plates, even at well sub-critical frequencies: hence u_n always has a component in phase with p, albeit small. The errors expressed by eqns (5.44) will produce an underestimate of the true radiated power, which will be most severe when the true value is very small, and is therefore not too serious in practice. On the other hand, the error in the estimate of u_n may significantly affect the accuracy of estimates of radiation efficiency of the vibrating surface (see Section 10.5).

In most practical cases, the detailed structure of the pressure field is not known *a priori*: near to complex extended sources it cannot even be guessed at. Hence it is in principle impossible to evaluate the magnitudes of the inherent errors associated with any one measurement. Measurements at a point with different microphone separations may give some indication of the sensitivity of the estimate to this parameter, and of the frequency range in which errors are negligible. However, most users do not have the time, patience or money to indulge in such a 'luxury'. Except in near fields, the plane wave errors given by eqns (5.40) are conservative, because the probe axis is not usually exactly coincident with the direction of the mean intensity vector; therefore $(kh)_{\text{axis}} < kh$.

5.6.2 Systematic errors inherent in the p-u technique

The most widely used p-u probe utilises a particle velocity transduction principle based upon the convection of an ultrasonic beam by the audio-frequency particle flow. A pair of oppositely directed beams are launched from the emitting transducers and received by another pair at a distance of about 28 mm. A beam pair is necessary to suppress the effect of audio-frequency temperature fluctuations associated with the sound field being measured. For the purpose of estimating the errors associated with the finite beam path length, only the distance between the emitting and receiving transducers needs to be known: we shall denote it by d.

At some point x along the beam path, the local speed of propagation of the ultrasonic wave, relative to earth, is given by

$$v(x, t) = c(x, t) \pm u(x, t) \tag{5.45}$$

where c is the speed of sound and u is the audio-frequency particle speed. The ultrasonic wave is considered to be convected by the audio-frequency wave, rather than vice versa, because the wavelength of the latter is at least twenty times that of the former. The time taken for an ultrasonic disturbance to traverse the elemental distance δx is given by

$$\delta t = \delta x / (c \pm u) \tag{5.46a}$$

Since $u \ll c$, this may be written as

$$\delta t \approx (\delta x / c)(1 \mp u/c) \tag{5.46b}$$

The acoustic phase speed c is a function of the absolute temperature of the air, which changes with acoustic pressure. In a plane progressive wave $\partial c / \partial u = (\gamma - 1)/2$ which produces an increment in c equal to $0 \cdot 2u$. Hence eqn (5.40) may be written as

$$\delta t^+ \approx (\delta x / c_0)(1 - 1 \cdot 2u/c_0) \tag{5.47a}$$

or

$$\delta t^- \approx (\delta x / c_0)(1 + 0 \cdot 8u/c_0) \tag{5.47b}$$

where the subscript 0 indicates the equilibrium condition.

Integration of this equation to obtain the transit time of an ultrasonic disturbance is not straightforward because the acoustic particle velocity u at any *position* x depends upon the time at which it is evaluated, and that relevant time depends upon how long it took the ultrasonic disturbance which is being followed to get there; this, in turn, depends upon the past history of the

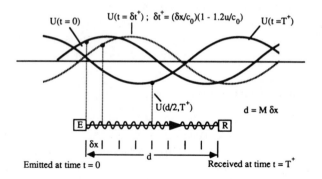

Fig. 5.3. Illustration of the state of the particle velocity fields at various times.

convection process. The process may be envisaged by reference to Fig. 5.3, which represents the beam path divided up into M small finite intervals δx.

For the ultrasonic wave travelling in the positive x-direction, we may consider the transit times for an ultrasonic disturbance as the sum of M elemental transit times, thus:

$$\delta t_1^+ = (\delta x/c)[1 - 1\cdot 2u(0,0)/c]$$
$$\delta t_2^+ = (\delta x/c)[1 - 1\cdot 2u(\delta x, \delta t_1^+)/c]$$
$$\delta t_3^+ = (\delta x/c)[1 - 1\cdot 2u(2\delta x, \delta t_1^+ + \delta t_2^+)/c]$$
$$\vdots \qquad \vdots \qquad \vdots$$
$$dt_N^+ = (\delta x/x)\left[1 - 1\cdot 2u\left((N-1)\delta x, \sum_{n=0}^{N-1} \delta t_n^+\right)\bigg/ c\right] \qquad (5.48)$$

The total transit time T^+ over a distance d is the sum of the incremental terms:

$$T^+ = \sum_{N=1}^{M} \delta t_N^+ = (\delta x/c) \sum_{N=1}^{M} \left[1 - 1\cdot 2u\left((N-1)\delta x, \sum_{n=0}^{N-1} \delta t_n^+\right)\bigg/ c\right] \qquad (5.49)$$

The ultrasonic disturbance will arrive at the receiver at time T^+ after the instant of emission, which will be defined as $t = 0$. Since the particle velocity is evaluated from the phase difference between the two ultrasonic signals, we must trace the progress of the negative-going ultrasonic wave from reception at time T^+ to the time of emission. Working back from the receiver to the emitter we get

$$\delta t_1^- = (\delta x/c)[1 + 0\cdot 8u(\delta x, T^+/c)]$$
$$\delta t_2^- = (\delta x/c)[1 + 0\cdot 8u(2\delta x, T^+ - \delta t_1^-)/c]$$
$$\vdots \qquad \vdots \qquad \vdots$$
$$\delta t_N^- = (\delta x/c)\left[1 + 0\cdot 8u\left(N\delta x, T^+ - \sum_{n=0}^{N-1} \delta t_n^-\right)\bigg/ c\right] \qquad (5.50)$$

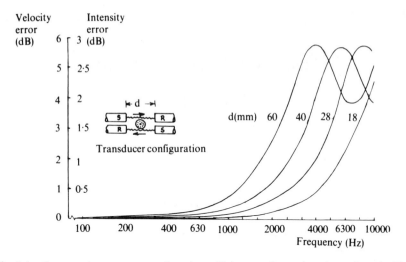

Fig. 5.4.   Systematic p-u error as a function of frequency for various beam lengths (*d*).

and

$$T^- = \sum_{N=1}^{M} \delta t_N^- = (\delta x/c) \sum_{N=1}^{M} \left[1 + 0\cdot 8u\left(N\delta x, \sum_{n=0}^{N-1} \delta t_n^-\right)\middle/c\right] \quad (5.51)$$

In the presence of a uniform convective field of speed *u*, the difference in phases between the received signals of frequency ω_u is

$$\delta\phi_0 = \delta\phi^- - \delta\phi^+ = \omega_u[T^- - T^+]$$
$$= \omega_u[d/(c-u) - d/(c+u)] = 2\omega_u ud/(c^2+u^2)$$
$$\approx 2\omega_u ud/c^2 \quad u \ll c \quad (5.52)$$

Where the convective field is a plane progressive sound wave, and the pressure signal is electronically delayed with respect to the particle velocity signal by a time *d*/2*c*, the appropriate particle speed is *u* at the acoustic centre of the microphone ($x = d/2$), at time $t = T^+/2$, i.e. $u = u(T^+/2, d/2)$. Hence a normalised error in $\delta\phi_0$, or *u*, may be evaluated as

$$e(u) = (c^2/2d)[(T^- - T^+)/u(T^+/2, d/2)] - 1 \quad (5.53)$$

This error has been evaluated numerically for a range of frequencies and ultrasonic beam lengths in air. The logarithmic form of *e*(*u*) is plotted in Fig. 5.4. Assuming perfect transduction by the associated pressure transducer, the dB error in estimated intensity is half that in estimated velocity.

A change of mean absolute temperature from T_0 to $T_0 + \delta T_0$ increases c_0 by a factor $[1 + \delta T_0/T_0]^{1/2}$. Since the measured phase change is inversely proportional to c_0^2, the sensitivity will change by a factor $[1 - 2\delta T_0/T_0]$.

CHAPTER SIX

Instrumentation and calibration

6.1 INTRODUCTION

The subjects of this chapter are the hardware used to measure sound intensity, associated signal processing procedures, the calibration and field checks of measurement systems and the effects of environmental factors. The scope of the description of the hardware is limited to general accounts of the transducer arrangements and signal processing systems which are currently in use at the time of revision of this book. For detailed descriptions and specifications of commercially available equipment the reader should contact equipment suppliers.

The rationale for the principal requirements of International Standard IEC 1043 which specifies the requirements for intensity measurement systems is presented. The chapter also contains a brief introduction to the effect of inter-channel phase response mismatch, a principal source of systematic error which is dealt with more comprehensively in Chapter 7.

6.2 GEOMETRIC ASPECTS OF SOUND INTENSITY PROBES

6.2.1 p-p Probes

Two nominally identical, high quality sound pressure transducers (normally condenser microphones or hydrophones) are placed close together in a support system which is designed to minimise the diffraction of the incident sound field. Most condenser microphone capsules take the form of short cylinders which may be associated in various configurations, including 'side-by-side', 'face-to-face', 'tandem' and 'back-to-back' (see Fig. 5.1). Naturally, the capsules have to be mounted on a support, which normally takes the form of a co-axial cylindrical body, or two separate such bodies.

A fundamental probe design parameter is the selection of the separation distance d appropriate to the design frequency range, remembering that any particle velocity signal is proportional to d, and that errors due to phase mismatch between transducer channels increase in severity as d is decreased: d should therefore be made as large as possible, consistent with acceptable inherent finite difference errors. The plane wave finite difference error is usually taken as a reference value, but, as shown in the previous chapter, it can be exceeded in reactive fields. For example, for a maximum plane wave mean intensity error of $-1\,dB$, $kd < 1\cdot2$, or $fd < 68$ in air at $20\,°C$. This is generally too large to be acceptable, because additional measurement errors occur, and it is preferable to restrict kd to less than $0\cdot8$ ($fd < 43$). This limit, which corresponds to a systematic error of $-0\cdot5\,dB$, is plotted against frequency in Fig. 6.1.

It is clearly preferable to use microphones with as high a sensitivity as possible, provided that they are extremely stable against variations of temperature and humidity. But sensitivity increases with area of the sensing element, thereby creating, in the case of the side-by-side arrangement, a conflict between this requirement and the restriction on kd. For example, the minimum d with half-inch diameter microphones is 13 mm, which, according to the restrictions stated above, dictates an upper usable frequency limit of about 3 kHz. In fact, diffraction errors also limit this configuration to about the same upper frequency; these are highest for sound waves approaching from the hemisphere of space which contains the microphone supports, as illustrated by Fig. 6.2.

A configuration which overcomes the separation limit, and also produces a calculable axi-symmetric diffraction field, is the face-to-face arrangement. It is necessary to place a solid plug in the space between the sensing elements in order to control the acoustic separation distance, which can be significantly different from the geometric distance. A commercial example is shown in Fig. 6.3. Comprehensive details of its diffraction behaviour are available in Ref. 54. The microphones may also be accommodated in a cylindrical body, as

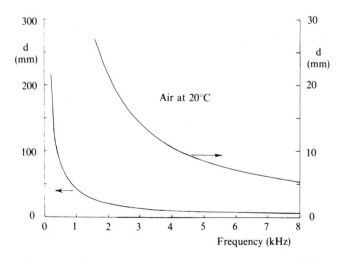

Fig. 6.1. Maximum p-p separation distance for a maximum systematic intensity error of $-0\cdot5\,dB$ as a function of frequency.

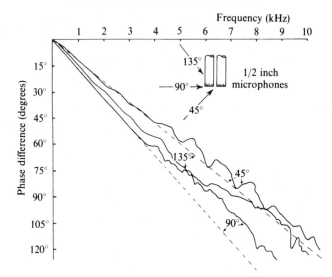

Fig. 6.2.   Diffraction effects of a side-by-side p-p configuration.

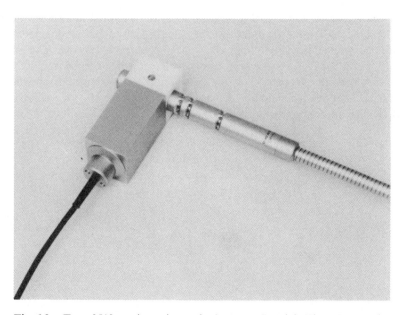

Fig. 6.3.   Type 3519 p-p intensity probe (courtesy Brüel & Kjaer, Denmark).

in the commercial product shown in Fig. 6.4. The back-to-back arrangement is also satisfactory, provided that the geometric separation distance is not much less than the diameter, i.e. the transducer assembly does not take the form of a disc.[55]

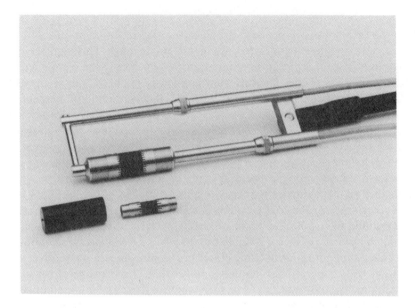

Fig. 6.4. Type CM-404 p-p sound intensity probe (courtesy Ono-Sokki, Japan).

In spite of the acoustic disadvantages of the side-by-side configuration, it has a number of mechanical advantages. It may be placed very close to radiating surfaces; it may be easily mounted so that it can be rotated about its axis of symmetry for field performance checks; and it may receive a standard plastic ball windscreen, which is necessary for outdoor use, and also provides valuable protection from mechanical and thermal damage.

It should be carefully noted that the existence of pressure equalisation vents in condenser microphones can produce low frequency errors in intensity measurements in fields with high spatial gradients of pressure amplitude or phase (such as standing waves or source near fields). This is because the vent samples the field at a different point from the diaphragm. Recently developed intensity microphones have vents with greatly improved low pass filter characteristics.[56]

It is the scattering effects of 'home made' microphone support systems which commonly constitute the dominant source of diffraction error. The sensitivity of intensity measurements to even apparently negligible reflections was first brought home to me when I became puzzled by a 'glitch' at 2 kHz in the inter-microphone phase difference of an intensity probe in a large anechoic chamber. The culprit was ultimately found to be a 2 cm diameter pole which supported the cotton line used to define the reference axis—the pole was 4 m distant from the probe! A review of the influence of microphone support systems is presented in Ref. 57.

6.2.2   Multi-microphone probes

The two-microphone p-p probe provides the signals to compute only the component of intensity directed along the probe axis. This limitation is acceptable for sound power determinations in which only the component of intensity normal to the measurement surface is required, but it presents a logistical problem for users who wish to establish a vector map of an intensity field, for the purpose of source location or the study of room acoustic energy flux, for example. Examples of multi-microphone probes are shown in Figs 6.5 and 6.6. The signals are combined in various ways depending upon the topology of the microphone array. Naturally, the increased bulk of solid material produces more severe diffraction effects than a p-p probe, which limits the useful maximum frequency to about 5000 Hz.

An alternative probe configuration in which twenty or more microphones are mounted in a spherical array provides partial compensation for phase mismatch by virtue of sensor redundancy (Fig. 6.7).[58,59]. It has the advantages over other configurations of being closer to an omni-directional receiver and having a wider operational frequency range than a p-p probe of the same overall dimension. It remains to be seen whether these advantages outweigh the increased requirement for signal conditioning and acquisition channels and the problem of developing a precision calibration technique.

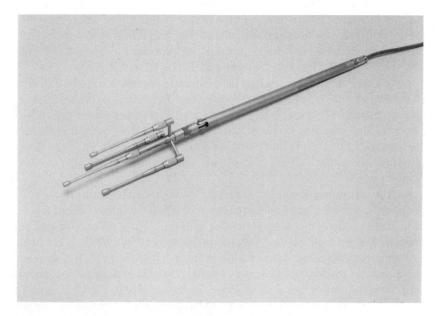

Fig 6.5   3–D Sound intensity microphone probe (Model MI–6420). (Courtesy of Ono Sokki, Tokyo, Japan.).

Fig. 6.6. A three-dimensional intensity probe (courtesy of Brüel & Kjaer, Denmark).

6.2.3 p-u Probes

The principal geometric requirements facing the designer of a p-u probe are (i) to produce effective coincidence of the points at which the sound pressure and particle velocity are sensed, and (ii) to minimise the diffraction effects of the probe assembly. The only widely used probe is shown in Fig. 6.8, in which the ultrasonic beam is arranged to pass within 3 mm of the acoustic centre of the microphone. Typical directional response is illustrated by Fig. 6.9.

In Section 5.6.2 the error in intensity estimate due to finite ultrasonic beam length was shown to limit the usable frequency range $(10 \lg[1 + e(I)] < 0.5 \, \text{dB})$ to $kd < 1.2$.

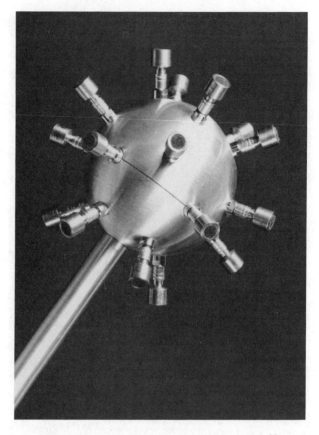

Fig. 6.7. A multi-microphone intensity probe.[58]

6.3 TRANSDUCER CHANNEL MISMATCH

6.3.1 p-p Probes

The matching of the impulse responses (or amplitude and phase responses) of the two transducers in a p-p probe is clearly of vital importance. The effect of phase response mismatch upon the accuracy of any particular measurement depends upon the relative magnitudes of the phase mismatch of the measurement system and the actual phase difference of the sound pressures at the transducer sensing points: the latter depends upon the nature of the sound field, and the location and orientation of the probe within the field. For example, in a plane progressive sound wave, the actual phase difference at the sensing points of a p-p probe varies from kd to 0 as the probe axis is rotated through 90° from an initial orientation along the direction of wave propagation. The associated fractional error due to phase mismatch varies from a finite value to infinity.

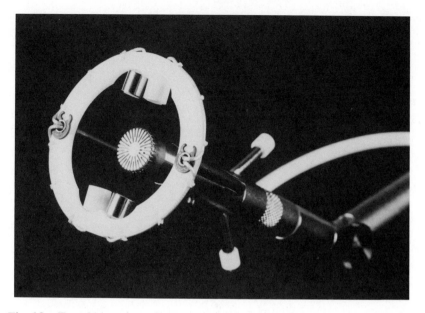

Fig. 6.8. Type 216 p-u intensity probe (courtesy Norwegian Electronics, Norway).

The phase difference kd in a plane progressive wave is conventionally taken as a convenient reference value for phase differences in other sound fields, although it is not the largest possible value. Figure 6.10 illustrates the general case in phasor form; the errors in the estimates of p and Δp are clearly seen. The actual pressure phase difference between the sensing points is ϕ_f, the transducer channel mismatch is $\pm \phi_s$, and the reference plane wave value is $\phi_0 = kd$. An approximate measure of the ratio $\phi_0/(\phi_f \pm \phi_s)$ may be obtained from eqn (5.7), which shows that $I \approx P_1 P_2 \phi/2\omega\rho_0 d$ when $\phi \ll 1$, which is always the case in the usable frequency range of a probe. In a plane progressive wave $\phi = \pm \phi_0 = \pm kd$, and therefore $I_0 = \pm P_1 P_2/2\rho_0 c$. Writing I_i as the indicated component of mean intensity in the direction of the probe axis, and $|I_i|$ as its absolute value

$$|I_i/I_0| = |(\phi_f \pm \phi_s)/\phi_0|$$

In logarithmic terms

$$10\lg|I_i/I_0| = 10\lg|I_i| - 10\lg[\overline{p^2}/\rho_0 c]$$
$$= 10\lg|I_i| - 10\lg[\overline{p^2}/p_{ref}^2] - 10\lg[p_{ref}^2/\rho_0 c]$$

Now, the reference value for I is $I_{ref} \approx p_{ref}^2/\rho_0 c$. Thus

$$10\lg|\phi_0/(\phi_f \pm \phi_s)| = 10\lg[\overline{p^2}/p_{ref}^2] - 10\lg|I_i/I_{ref}|$$
$$= L_p - L_{|Ii|}\,\mathrm{dB} \qquad (6.1)$$

The sign of the indicated linear intensity component I_i indicates the sign of

Pressure microphone: The response for three frequencies (20 Hz ····, 1 kHz — and 5 kHz ---) is indicated in the diagram below.

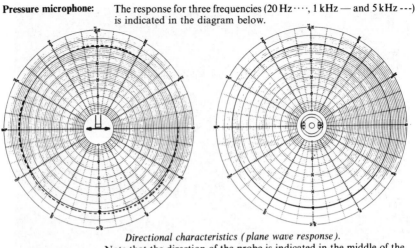

Directional characteristics (plane wave response).
Note that the direction of the probe is indicated in the middle of the circular diagram.

Velocity microphone: The response for three frequencies (20 Hz ····, 1 kHz — and 5 kHz ---) is indicated in the diagram below. .

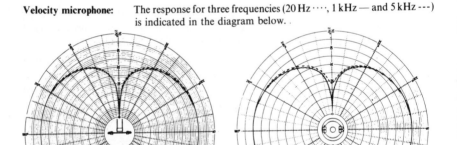

Directional characteristics (plane wave response).
Note that the direction of the probe is indicated in the middle of the circular diagram.

Intensity probe: The directional characteristics of the intensity are equal to the velocity characteristics except for the level change indicated as 10 dB in the velocity diagram which equals a level change of 5 dB in the intensity.

Fig. 6.9. Directional response of the Type 216 intensity probe (courtesy Norwegian Electronics, Norway).

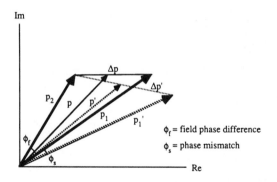

Fig. 6.10. Errors in estimates of pressure and pressure difference due to phase mismatch.

$\phi_f \pm \phi_s$. The difference betwen the indicated values of Sound Pressure Level and Sound Intensity Component Level is known as the 'Pressure-Intensity Index', denoted by δ_{pI} introduced in Section 4.6.3:

$$\delta_{pI} = L_p - L_{|Ii|} \, dB \qquad (6.2)$$

(The term 'Sound Intensity *Component* Level' is used because the probe indicates only the component of the intensity in the direction of the probe measurement axis.)

At the risk of being repetitive, it is emphasised yet again that δ_{pI} is a compound function of (i) the form of sound field; (ii) the position and orientation of the probe in that sound field; and (iii) the transducer channel mismatch. It is not very sensitive to transducer separation distance d, unless ϕ_s is of a similar value to ϕ_f, which is clearly not acceptable if left uncorrected. Since phase mismatch causes an indicated value of δ_{pI} to be different from the true field value, by introducing an error into indicated intensity level, δ_{pI} itself cannot strictly be an indicator of error due to that same phase mismatch. However, provided the phase mismatch is much less than the field phase difference, δ_{pI} forms a useful guide to the 'difficulty' of making an accurate measurement of intensity, as we shall now see.

The normalised systematic error in indicated intensity due to phase mismatch may be written as $e_\phi(I) = \phi_s/\phi_f$. If the probe is placed in a specially controlled sound field of uniform pressure, in which $\phi_f = 0$, and $I_i = 0$, the ratio $\phi_0/(\phi_f \pm \phi_s)$ becomes equal to $\pm \phi_0/\phi_s$: the corresponding pressure-intensity index is known as the 'Pressure-Residual Intensity Index', denoted by δ_{pI0}. Hence δ_{pI0} is a measure of the phase mismatch of the measurement system, and is therefore a quality indicator. The difference between δ_{pI0} and δ_{pI} measured in a sound field is a measure of the fractional error $e_\phi(I)$, provided that ϕ_s is independent of the form of the sound field.

$$\delta_{pI0} - \delta_{pI} = 10\lg|1 + \phi_f/\phi_s| = 10\lg|1 + (1/e_\phi(I))| \, dB \qquad (6.3)$$

For a normalised error of ± 0.25, or an error in the estimate of I of ± 1 dB,

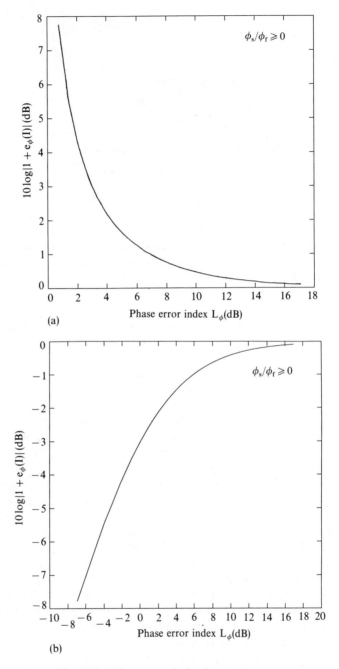

Fig. 6.11.   Phase error index for p-p systems.

$\delta_{\text{pI0}} - \delta_{\text{pI}} \approx 7\,\text{dB}$. As an indication of the significance of phase mismatch, this quantity might well be termed the 'Phase Error Index', although, unfortunately, the error increases as this 'index' decreases. For the sake of concision, it will henceforth be denoted by L_ϕ in this book. In Fig. 6.11, $e_\phi(I)$, and the equivalent value of $10\lg|1 + e_\phi(I)|$, are plotted against L_ϕ. Procedures for the experimental determination of δ_{pI0} are described in Section 6.9 on calibration. A typical spectrum of δ_{pI0} is shown in Fig. 6.12.

In addition to phase mismatch, the two transducers may differ in sensitivity. The effect is illustrated in Fig. 6.13, in which the prime on p_2' indicates that the sensitivity of transducer 2 exceeds that of its colleague. The pressure sum and difference are altered in magnitude and phase by the sensitivity mismatch; so, therefore, are the estimated pressure and particle velocity. The effect on the estimate of mean intensity in the frequency domain is most easily appreciated by reference to the harmonic form $I = P_1 P_2 \sin\phi/2\rho_0\omega d$, in which

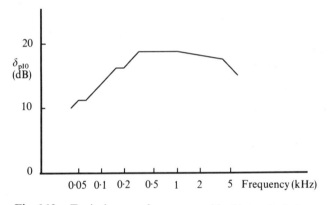

Fig. 6.12. Typical curve of pressure-residual intensity index.

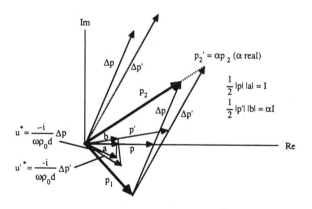

Fig. 6.13. Phasor diagram showing the effect of p-p transducer sensitivity mismatch $(1:\alpha)$.

$\phi = \phi_1 - \phi_2$. In the absence of coexisting phase mismatch, sensitivity mismatch produces no error in the estimate of mean intensity, provided the individual sensitivities are accurately known. Naturally, the estimate of reactive intensity, which is proportional to the difference of mean square pressures, is highly sensitive to errors in individual transducer channel calibration. Estimates of both the amplitude and phase of particle velocity, which is proportional to pressure difference, are also sensitive to such errors, which will also affect the accuracy of estimates of time-dependent intensity in non-stationary fields (eqn (5.5)). If phase mismatch is present, the effects of both types of mismatch become compounded in a most complicated fashion, the resultant error depending on the characteristics of the field under investigation.

Phase mismatch also distorts the directional sensitivity of a p-p probe. Of particular importance is the fact that the null in intensity response, which is often used to locate concentrated source regions, is altered from the 90° direction to one given by $\phi_m = \cos^{-1}(\phi_s/kd)$, or, expressed in terms of deviation from the ideal direction, $\phi_m = \sin^{-1}(-\phi_s/kd) \approx -\phi_s/kd$. Hence, the effect of a given phase mismatch decreases with increasing frequency. The difference between on-axis sound intensity levels indicated in the 'forward' and 'reverse' positions is given by

$$\delta L_1 = 10 \lg [(1 + B)/(1 - B)] \, dB \qquad (6.4)$$

where $B = \phi_s/kd$.[60] This formula provides a rapid, approximate means of estimating ϕ_s from probe reversal in any field.

Phase mismatch error is treated in more detail in Chapter 7.

6.3.2   p-u Probes

The phase response matching problem of p-u probes is created essentially by the combination of transducers which are based upon entirely different principles and mechanisms of transduction. A typical microphone phase response curve is shown in Fig. 6.14. In well-matched p-p probes, the two curves rise almost identically with frequency, but the corresponding phase response curve for an ultrasonic particle velocity transducer lies close to zero degrees over the entire operational frequency range. Consequently, the difference between the p and u phase responses has to be compensated by electronic means, which is not an easy task with analogue circuitry if the curve is irregular: it could, however, be readily accomplished by fairly high order digital filters. Each p-u pair produced has to be individually compensated.

The significance of p-u phase mismatch, which is illustrated in Fig. 6.15, is related to the form of the acoustic field being measured, and to the probe orientation, as with the p-p probe, but the expressions take a different form. In simple harmonic terms, the indicated mean intensity $I_i = \frac{1}{2} PU \cos(\phi_f \pm \phi_s)$ and the true intensity $I = \frac{1}{2} PU \cos(\phi_f)$, where now the angles ϕ_f and ϕ_s denote the field angle between pressure and particle velocity, and the instrumentation

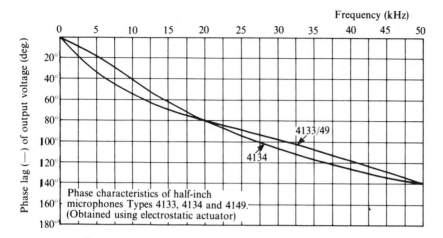

Fig. 6.14. Typical pressure phase responses for 0·5 in condenser microphones (courtesy Brüel & Kjaer, Denmark).

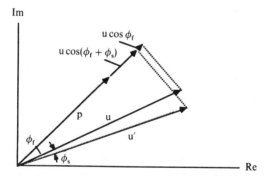

Fig. 6.15. Effects of p-u phase mismatch.

phase mismatch, respectively. The normalised error is given by

$$e_\phi(I) = [\cos(\phi_f \pm \phi_s)/\cos(\phi_f)] - 1 = \cos(\phi_s) \mp \tan(\phi_f)\sin(\phi_s) - 1 \quad (6.5)$$

When $\phi_s \ll 1$

$$e_\phi(I) \approx \phi_s \tan \phi_f \quad (6.6)$$

Values of $10\lg[1 + e_\phi(I)]$ are plotted against ϕ_f for fixed values of ϕ_s in Fig. 6.16. Hence the significance of a phase error ϕ_s *increases* with the field phase difference. At first sight, this behaviour seems contrary to that of the p-p probe, until it is realised that, for a given pressure, small p-p phase angles correspond to low intensities, whereas small p-u phase angles correspond to high intensities. Equations (6.5) and (6.6) indicate that the normalised error $e_\phi(I)$ will normally be greatest when $\phi_f \to \pi/2$. The pressure-intensity index is

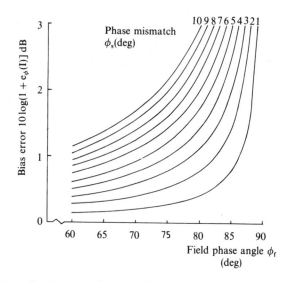

Fig. 6.16. Normalised systematic error of a p-u system due to phase mismatch as a function of field phase angle.

given by

$$\delta_{\mathrm{pI}} = 10 \lg (|z|/\rho_0 c) + 10 \lg [\sec(\phi_\mathrm{f} \pm \phi_\mathrm{s})] \, \mathrm{dB} \tag{6.7}$$

where z is the complex specific acoustic impedance at the measurement point in the direction of the probe axis; the magnitude of z can take values between zero and infinity. Hence, unlike the case of the p-p probe, there is no unique relationship between δ_{pI} and the normalised phase mismatch; therefore the associated bias error cannot be estimated from the measured value of δ_{pI} alone.

When a p-u probe is placed with its axis co-linear with the direction of propagation of a plane progressive wave

$$\delta_{\mathrm{pI}} = 10 \lg [\sec(\phi_\mathrm{s})] \, \mathrm{dB} \tag{6.8}$$

Unlike δ_{pI0} for the p-p probe, this is not a particularly useful quantity because $\phi_\mathrm{s} \ll 1$ and therefore it always takes values close to zero. As indicated by eqn (6.6), error in the intensity estimate due to phase mismatch is greatest when $\phi_\mathrm{f} \gg \phi_\mathrm{s}$, not when $\phi_\mathrm{f} \rightarrow 0$. In this case we may approximate eqn (6.7) by

$$\delta_{\mathrm{pI}} \approx 10 \lg (|z|/\rho_0 c) + 10 \lg [\sec(\phi_\mathrm{f})] \, \mathrm{dB} \tag{6.9}$$

Equation (6.6) suggests that the appropriate field in which to check the phase mismatch of p-u systems should be strongly reactive, such as that in a standing wave tube with a reflective termination. As indicated by Fig. 6.14, phase mismatch may occur in any part of the frequency range in which compensation is inadequate. This creates an experimental problem which has not to date been fully resolved. The major practical difficulty is that the physical size of a p-u probe prohibits its installation in the small diameter tube necessary to generate a well-controlled one-dimensional, high frequency standing wave.

Let us assume that a p-u probe is placed in such a field in which the pressure is given by

$$p(x,t) = A[\exp(-ikx) + R\exp(i\theta)\exp(ikx)]\exp(i\omega t) \qquad (6.10)$$

The Standing Wave Ratio is defined as

$$SWR = 20\lg[(1+R)/(1-R)]\,dB \qquad (6.11)$$

The particle velocity is given by

$$u(x,t) = (A/\rho_0 c)[\exp(-ikx) - R\exp(i\theta)\exp(ikx)]\exp(i\omega t) \qquad (6.12)$$

and the mean intensity is given by eqn (4.24) as

$$I = (A^2/2\rho_0 c)[1 - R^2] \qquad (6.13)$$

If the phase mismatch is ϕ_s (radians), the indicated particle velocity may be written as

$$u'(x,t) = u(x,t)\exp(i\phi_s)$$

and the indicated intensity is

$$I_i = (A^2/2\rho_0 c)[(1 - R^2)\cos(\phi_s) + 2R\sin(\theta + 2kx)\sin(\phi_s)] \qquad (6.14)$$

The magnitude of the error is seen to depend upon the magnitude and phase of the complex reflection coefficient $R\exp(i\theta)$, on the position of the probe in the standing wave, and on the phase mismatch ϕ_s. The latter may be estimated by placing a probe in a standing wave tube, evaluating the complex reflection coefficient, and the wave amplitude A in the usual way, and then determining ϕ_s from eqn (6.14). The largest effects of phase mismatch will be observed at positions halfway between pressure maxima and minima, where

$$e_\phi(I) \approx 2R\phi_s/(1 - R^2) \qquad (6.15)$$

6.4 THE EFFECTS OF MEAN AIRFLOW, TURBULENCE AND WINDSCREENS ON PROBE PERFORMANCE

These effects are more fully dealt with in Chapter 11. However, a few observations specifically concerning probe design and operation are appropriate at this point.

6.4.1 p-p Probes

Strictly speaking, the p-p principle is invalid in the presence of mean flow, however small, because it is based upon the zero mean flow momentum equation (3.12). For example, eqn (3.8) shows that in a one-dimensional plane sound wave travelling in a uniform medium, which is itself flowing uniformly in the same direction at speed U, the total fluid particle acceleration is given by

$$Du/Dt = \partial u/\partial t + u\partial u/\partial x \qquad (6.16)$$

where $u = U + u'$, and u' is the acoustic component of the total particle velocity. Hence

$$Du'/Dt = \partial u'/\partial t + (U + u')\partial u'/\partial x \qquad (6.17)$$

Let $u'(x, t) = A \exp[i(\omega t - kx)]$. Then

$$Du'/Dt = i\omega u' - iku'(U + u')$$
$$= i\omega u'[1 - U/c - u'/c]$$
$$\approx i\omega u'[1 - M] \qquad (6.18a)$$

where M is the mean flow Mach number. The momentum equation becomes

$$\partial p/\partial x = -i\omega\rho_0 u'[1 - M] \qquad (6.18b)$$

Hence eqn (5.1) is not valid. Although the error would appear to be small if $M \ll 1$, this is not necessarily so in strongly reactive fields, as we shall see in Chapter 11.

There will always be unsteady components in any airflow encountered in practical intensity measurements. These may exist in the form of turbulence in the oncoming flow, such as wind, or flow in a ventilation duct, or in turbulence created by the presence of the probe itself. The adverse effects of turbulence on the accuracy of low frequency ($< 200\,\text{Hz}$) intensity measurements are generally more serious than errors incurred by neglecting the effects of mean flow convection. The transducers in a p-p probe cannot distinguish acoustic pressures (associated with fluid compressibility and the propagation of energy at the speed of sound) from unsteady hydrodynamic pressure fluctuations (associated with local 'incompressible' momentum fluctuations). In addition, the latter can overload the signal conditioning system, thereby adversely affecting the signal-to-noise ratio. Therefore the transducers must be protected from direct exposure to turbulence by enclosure in a windscreen, the porous structure of which offers resistance to impinging flow. A windscreen also offers protection of the probe against mechanical and thermal damage. An example of a commercial, open cell plastic foam windscreen, is shown in Fig. 6.17.

In principle, a p-p probe possesses a degree of in-built capability for rejection of turbulence pressures because these are well correlated over regions of space which are generally small compared with an acoustic wavelength. The time-average product operation on the two sensed pressures, which is implicit in all p-p measurements of mean intensity, excludes components of a turbulent field which are correlated over distances small compared with the transducer separation distance: random errors are increased by coherence loss. In practice, the resulting benefits are rather small in free turbulent flows, especially at low frequencies: only probes with separation distances in excess of 50 mm provide any significant suppression by this mechanism.

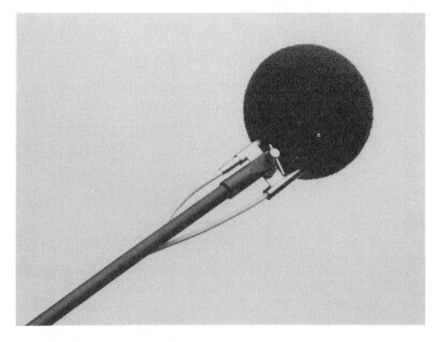

Fig. 6.17. Porous foam windscreen (courtesy Brüel & Kjaer, Denmark).

A small-scale study of the influence of low-speed airflow on measurements of sound intensity, and on the associated estimates of source sound power, has been made by Jacobsen.[61] Figures 6.18 and 6.19 present comparisons of estimates of sound intensity and sound pressure levels made with and without a windscreen over the probe and for two different microphone separations. The principal conclusions from the study are as follows: unsteady airflow produces false particle velocity, intensity and sound pressure signals, in decreasing order of severity, particularly at frequencies below about 200 Hz; the use of a windscreen greatly reduces these false signals and the resulting false intensity may be reduced by increasing the microphone separation; the opposition of false (hydrodynamic) and true (acoustic) intensity vectors produces enormous random errors as well as bias error. The establishment of reliable guidelines for good practice is contingent upon further studies of the effects of turbulent flow on sound intensity measurement in which the influence of turbulence intensity and spatial scales are determined.

The presence of a foam windscreen which presents a resistance to both mean and oscillatory airflow alters the acoustic particle velocity from its 'free' value. The effect on intensity of a foam ball of the type shown in Fig. 6.17 is to alter it by a fractional amount given by $(r/\omega\rho_0)(J/I)$, where r is the foam resistivity and J/I is the ratio of reactive to active intensity.[62] In practice, this error will only be significant at low frequencies in highly reactive fields, such as strong standing waves in enclosures, near inefficiently radiating loudspeakers at low

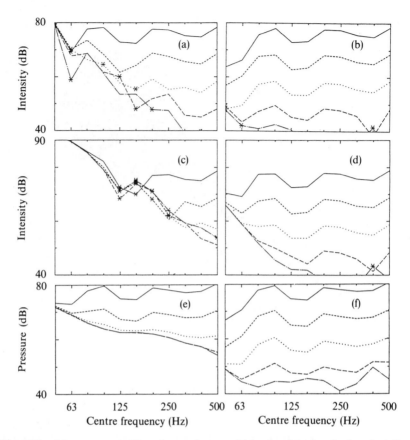

Fig. 6.18. Measurements 60 cm from a fan on which a loudspeaker is placed; airflow speed about $4.2\,\mathrm{m\,s^{-1}}$: (a), (b) 50 mm spacer; (c)–(f) 12 mm spacer; (a), (c), (e) no windscreen; (b), (d), (f) windscreen. Nominal sound power of source: —— 100 dB; ------ 90 dB; .... 80 dB; ––– 70 dB; –·–· 60 dB. * = Negative estimate.[61]

frequency and thin vibrating plates,[63] but it can be very large and even be larger than unity so that the intensity estimate has the wrong sign. Figures 6.20 and 6.21 illustrate the effect.

6.4.2 p-u Probes

The principle of ultrasonic transduction of particle velocity is not fundamentally altered by the presence of airflow; d.c. signals due to mean flow may easily be filtered out. Unfortunately, unsteady velocity components due to turbulence produce signals which cannot be distinguished from, and are usually much larger than, those due to acoustic waves. As with the p-p probe, a windscreen must be employed to shield the ultrasonic beams from the hydrodynamic field: a porous plastic windscreen serves also to suppress stray

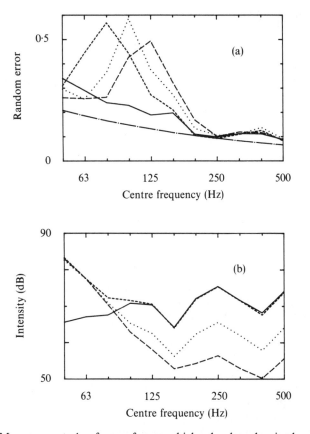

Fig. 6.19. Measurements 1 m from a fan on which a loudspeaker is placed; airflow speed about $5.5 \, \text{m s}^{-1}$, 12 mm spacer, windscreen: (a) normalised random error with an averaging time of 2 s; $-\cdot-$ theoretical minimum value, $1/(BT)^{1/2}$; (b) measured intensity. Nominal sound power of source: ------- 100 dB; .... 90 dB; $---$ 80 dB; —— 100 dB, no flow.[61]

ultrasonic wave reflections from nearby objects. There is some evidence to suggest that this form of p-u probe is more sensitive to unsteady flow than the p-p system.

6.5 ENVIRONMENTAL EFFECTS ON PROBE PERFORMANCE

Transducer performance is affected by ambient temperature, humidity, and various other phenomena of less practical significance, such as magnetic fields; it also varies with age, and is a complex function of environmental history. Readers are directed to manufacturers' data for detailed information, and only some of the more important general points will be mentioned here:

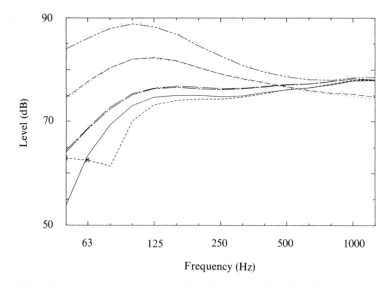

Fig. 6.20. Measurements made about 12 cm from a small enclosed loudspeaker: ——— active intensity, no windscreen; ----- active intensity, ellipsoidal windscreen; —·— pressure, no windscreen; – – pressure, ellipsoidal windscreen; ----- particle velocity, no windscreen; .... particle velocity, ellipsoidal windscreen; — — — reactive intensity, no windscreen; .... reactive intensity, ellipsoidal windscreen. * = negative estimate.[62]

(i) Only highly stable, high quality condenser microphones are suitable for intensity probes.

(ii) Temperature gradients in the vicinity of hot bodies, such as I.C. engine blocks, can produce temporary p-p mismatch. It is therefore advisable not to measure within regions of high temperature gradient.

(iii) The sensitivity of an ultrasonic p-u probe is directly proportional to absolute temperature, as shown in Section 5.6.2.

(iv) Significant changes can be produced in the phase response characteristics of microphones by mechanical shock, with negligible effect on sensitivity as checked, for example, by a pistonphone.

(v) Exposure of probes and associated cables to vibration may not only generate spurious signals by electrical means, but may modulate the particle velocity as sensed by the probe.

6.6   SIGNAL PROCESSING PROCEDURES

6.6.1   General implementation

The processing of p-p probe output signals to provide time-dependent intensity information can be performed only by the direct implementation of eqn (5.5) in the time domain, since $I(t)$ and $I(\omega)$ do not form a Fourier

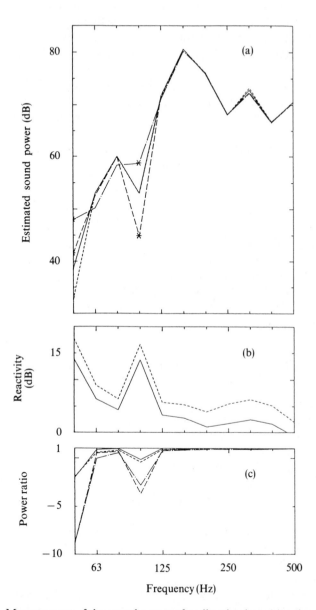

Fig. 6.21. Measurement of the sound power of a vibrating box: (a) estimated sound power —— 50 mm spacer, no windscreen; ---- 12 cm spacer, no windscreen; – – – 50 mm spacer, ellipsoidal windscreen; —·— 12 mm spacer, spherical windscreen; * = negative estimate; (b) error indicator: —— ellipsoidal windscreen; ----- spherical windscreen; (c) ratio of estimated to true sound powers: —— measured with ellipsoidal windscreen; ---- predicted with $\phi = 45\,\text{kg}\,\text{m}^{-3}\,\text{s}^{-1}$; – – – measured with spherical windscreen; predicted with $\phi = 62\,\text{kg}\,\text{m}^{-3}\,\text{s}^{-1.62}$

transform pair. Block diagrams of the analogue and digital signal processing systems are shown in Figs 6.22(a) and (b). Analogue or digital filtering may be applied identically to the two probe signals to produce frequency-band-limited time histories of the measured components of intensity.

For frequency domain analysis of stationary signals, either direct filtering may be applied, or Fourier transformation may be employed to evaluate the expressions in eqns (5.13) (p-u), or (5.19) (p-p), as appropriate. Block diagrams are shown in Figs 6.23(a) and (b). As indicated by eqn (5.19), it is not necessary to form the pressure sum and difference from p-p signals prior to spectral processing. The division by ω may be accomplished by post-processing, or alternatively, one of the two pressure signals may be integrated by the FFT analyser, in which case the expression for mean intensity component becomes

$$I_n(\omega) = \operatorname{Re} \{G_{p2z1}(\omega)\}/\rho_0 d \qquad (6.19)$$

where $z_1 = \int p_1 \, dt$. A disadvantage of this procedure is that the dynamic range of z_1 can be very large if a wide frequency range is analysed.

Spectral analysis may also be made of transient sound power flow, provided that the complete time histories of each signal are transformed using a rectangular time window, and that no spectral averaging is performed. In eqns (5.13) and (5.19) the power spectral densities are replaced by energy spectral densities. Further details and examples of this form of analysis are presented in Sections 5.4 and 10.6.

6.6.2 Practical advantages and disadvantages of direct and FFT intensity measurement systems

Intensity measurement instruments which implement the direct filtering principle normally output and display results in octave or fractional octave frequency bands, of which the bandwidth is proportional to the centre frequency. For example, the bandwidth of a one-third octave filter centred on 100 Hz is about 23 Hz. Instruments based upon FFT analysis generate spectral lines at *uniform* frequency intervals (frequency resolution) equal to the inverse of the length of one signal segment (a sample of signal time history used to produce one estimate of the signal spectrum). FFT analysers usually generate 400, 500, 800 or 1600 spectral lines. Consequently, the frequency resolution depends upon the bandwidth of the analysis selected by the user. For example, the choice of a baseband of 10 000 Hz on a 500 line analyser yields a frequency resolution of 20 Hz, and a corresponding signal segment length of 50 ms. Clearly, this resolution is entirely inadequate if it is intended to synthesise the one-third octave band spectrum level at 100 Hz from spectral line data; the FFT resolution is far too coarse because at least ten spectral lines should be used in the synthesis of each band. Consequently, the segment length must be increased by a factor of ten.

As we shall see in Section 7.3, the random error of spectral estimates is inversely proportional to $N^{1/2}$, where, in the case of direct filtering, N is the

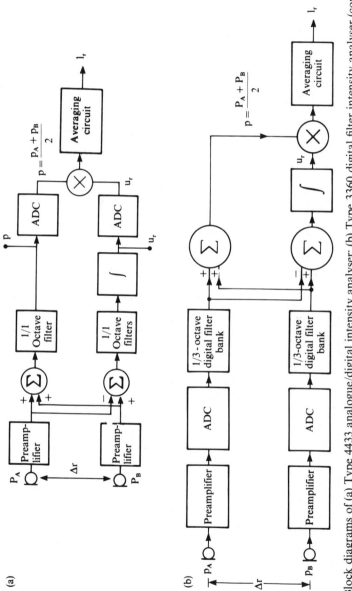

(a)

(b)

Fig. 6.22. Block diagrams of (a) Type 4433 analogue/digital intensity analyser; (b) Type 3360 digital filter intensity analyser (courtesy Brüel & Kjaer, Denmark).

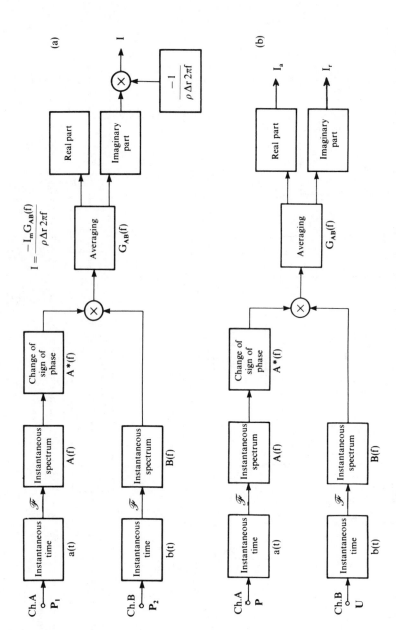

Fig. 6.23. Block diagrams of (a) FFT analysis for p-p systems; (b) FFT analysis for p-u systems (courtesy Brüel & Kjaer, Denmark).

product of the filter bandwidth and the total period of time allowed for the spectral estimation process. In the case of FFT analysis, N is simply the total number of *independent* signal segments used to form the spectral estimate from the ensemble of single segment transforms. (The 'independence' of signal segments is emphasised because FFT analysers use 'overlapped' segments to compensate for the loss of signal information that is caused by 'windowing' each segment to reduce the adverse effects of time truncation: spectral estimates from individual overlapped segments are clearly not independent. Finite processing time, plus the need to overlap, restrict the upper frequency for which 'real-time' operation is possible: 'real-time' processing means that *all* the digitised samples of the input signals are processed as they are acquired. If the frequency span of the analyser exceeds the real-time frequency, signals are not correctly processed, and the resulting spectral estimates may be seriously in error, especially in cases of periodic signals having high crest factors, such as those generated by cyclically operating machines.)

Returning to the above discussion of the generation of constant percentage bandwidth spectra by direct filter and FFT procedures, it is clear that the need for at least ten spectral lines in each band requires an FFT 'averaging' time ('real time' with no loss of data) at least ten times as long as the direct filter 'averaging time', if the random error in the estimate of *each* FFT spectral line is to equal that of the direct spectral estimate. However, since the random errors of the individual line estimates are nearly statistically independent, the random error of the estimate obtained by summing n individual line estimates is reduced by a factor $n^{1/2}$. Thus, in principle, the required averaging times for each procedure are equal.

In practice, the measurement times required when using FFT analysis to synthesise constant percentage bandwidth data are considerably longer than those for the direct filter technique because of the real-time limitations mentioned above, together with the fact that the synthesis computation also takes time. Where long averaging times are required because of adverse measurement environments, such as highly reverberant surroundings or strong extraneous noise sources, this factor can be of considerable importance.

Another disadvantage of FFT-based intensity analysis is that a display of many spectral lines, often containing complex mixtures of positive and negative values, is extremely difficult to interpret, and is almost useless for the purpose of making a quick assessment of the spatial distribution of intensity around a source because the rapid fluctuation of the many lines is impossible to monitor visually, and synthesised band data are always 'out-of-date'.

One very important advantage of most modern direct filter intensity analysers is that they can generate a particle velocity output signal, which may be used in measurements of acoustic impedance, radiation efficiency of vibrating surfaces, in teaching demonstrations, and in many other applications. FFT analysers can generate particle velocity information from p-p probe signals only indirectly.

Advantages of FFT analysers include the availability of coherence data which is useful in assessing random errors in intensity estimates, and in making intensity measurements in airflow, together with transfer functions which may be used to correct for inter-channel phase mismatch. Of course, the FFT analyser has a multitude of functions other than intensity measurement, and the intending purchaser of intensity measurement equipment must weigh up the pros and cons of the choice in the light of his or her other requirements. However, I must clearly state my preference for lightweight, battery-powered, direct filter systems for general purpose field survey and measurement, especially where dB(A) is of prime interest.

A useful combination for *in situ* impedance measurement is a direct system for generating pressure and particle velocity signals, plus an FFT analyser to generate the appropriate transfer function.

6.7   DIGITISATION, QUANTISATION AND DYNAMIC RANGE

It is vital that the two analogue signals from the intensity probe are synchronously sampled and held by an analogue-digital conversion (ADC) system. This condition is not always satisfied by multi-channel signal acquisition systems. The use of analogue tape recorders is not recommended; a far better solution is to employ digital audio equipment and video recorders.[64]

Digitisation produces uncertainties in intensity estimates which are related to quantisation resolution. For measurement of continuous signals with high crest factors, for example from cyclic impact machines, or from isolated transients, it is necessary to use at least 12-bit (0·025% full scale) ADC quantisation.[65] No general expressions for the dynamic range capability of an FFT-based intensity measurement system can be given because the accuracy with which the imaginary part of cross-spectral density is calculated depends upon the precise implementation of the arithmetic operations in the analyser. However, apart from the fairly obvious limitations imposed by electrical noise in the transducer signal conditioning circuits, the user of intensity measurement equipment should always give attention to possible dynamic range problems when attempting to analyse broad band signals of which the spectral levels vary widely over the frequency range of analysis. Intensity 'noise' floor levels should be established in pre-test laboratory checks, especially where measurements are to be made at low levels (e.g. flanking transmission in buildings).

A useful test of intensity analysers which operate with p-p probes is to feed a common pink noise signal to the two inputs, and to monitor the dB difference between indicated sound pressure level spectrum and indicated sound intensity level spectrum as the input level is decreased in, say, 10 dB steps. The effect of auto-ranging should also be investigated (see Section 6.8).

6.8 INTERNATIONAL STANDARD FOR SOUND INTENSITY INSTRUMENTS

International Standard IEC 1043: 'Electroacoustics—Instruments for the measurement of sound intensity—Measurement with pairs of pressure sensing microphones' was published in December 1993. The primary purpose of the Standard is to ensure the accuracy of sound intensity applied to the determination of sound power in accordance with International Standards ISO 9614-1 and 9614-2 (see Section 9.4). It covers the following matters: definitions; grades of accuracy; requirements for processors and probes; requirements for instruments which combine probes and processors; requirements for probe calibrators; processor, probe and calibrator performance verification procedures; field calibration and performance checks. The purpose of this section is not to reproduce the detail of the Standard but to explain the background to its principal requirements.

Instruments, processors and probes are classified as being class 1 or class 2 according to the measurement accuracy achieved. Separate requirements are specified for probes and processors because a user may combine these components from different suppliers to form an instrument. A class 1 instrument may only be formed by combining a class 1 probe with a class 1 processor. Class 1 processors shall, at least, cover the range from 45 Hz to 7·1 kHz in one-third octave bands. Class 2 processors shall, at least, cover the same range, or 45 Hz to 5·6 kHz in octave bands, as specified in ISO 9614.

Minimum required values of the pressure-residual intensity index δ_{pI0} are specified at one-third octave centre frequencies for probes, processors and instruments in Table 6.1. When the probe microphones are subjected to identical pressures in a residual intensity testing device the ratio of residual intensity to mean square pressure produced by a particular phase mismatch is inversely proportional to microphone separation distance d (eqn (7.4)): the tabulated values apply to a 25 mm microphone spacing and must be adjusted by a factor $+10\lg(d(\text{mm})/25)$ for other spacings. The effect of this requirement on δ_{pI0} is to limit the maximum allowable phase mismatch. For example, a specified value of δ_{pI0} of 20 dB corresponds to a phase mismatch of one-hundredth of the phase difference kd between the pressures at the sensor positions in an axially propagating plane progressive wave: at 1000 Hz, this corresponds to a phase mismatch of about 0·26°.

The measurement standards ISO 9614-1/2 define the dynamic capability L_d of an instrument as δ_{pI0} minus a bias error factor K which varies between 7 and 10 dB according to the grade of accuracy required of the sound power determination (see Section 9.4.1(a)). In order to restrict the phase mismatch bias error to a value compatible with the stated confidence limits on reproducibility, the field pressure-intensity index may not exceed $(\delta_{pI0} - K)$ dB. The requirements in IEC 1043 represent a compromise between current technical capabilities and the maximum likely demands presented by adverse measurement environments. However, there will be test environments which will

Table 6.1 Minimum pressure-residual intensity index requirements for probes, processors and instruments for the 25 mm nominal microphone separation in decibels

Band centre frequency (Hz)	Probe		Processor		Instrument	
	Class 1	Class 2	Class 1	Class 2	Class 1	Class 2
50	13	7	19	13	12	6
63	14	8	20	14	13	7
80	15	9	21	15	14	8
100	16	10	22	16	15	9
125	17	11	23	17	16	10
160	18	12	24	18	17	11
200	19	13	25	19	18	12
250	20	14	26	20	19	13
315	20	15	26	20	19	14
400	20	16	26	20	19	14·5
500	20	17	26	20	19	15
630	20	18	26	20	19	16
800	20	18	26	20	19	16
1000	20	18	26	20	19	16
1250	20	18	26	20	19	16
1600	20	18	26	20	19	16
2000	20	18	26	20	19	16
2500	20	18	26	20	19	16
3150	20	18	26	20	19	16
4000	20	18	26	20	19	16
5000	20	18	26	20	19	16
6300	20	18	26	20	19	16

Note: for pressure-residual intensity requirements for microphone separations other than 25 mm, add $10\lg(x/25)$, where x is the microphone separation in mm, to the figures in the table. (Reproduced with permission from IEC 1043 (1993).)

defeat the best current instruments. (Note: some instruments incorporate phase mismatch 'compensation' devices (see Section 7.2) but their efficacy is limited by instrument electrical noise: consequently their apparent values of δ_{pI0} may vary randomly from determination to determination.)

The frequency response of the probe is tested in a plane progressive wave by comparison with a microphone of known free field response in the range 500–6300 Hz. Various forms of test signal are permitted, including gated tone bursts to allow editorial excision of room reflections. According to Jarvis,[66] the use of continuous, wide band noise signals in a high performance anechoic chamber is the best option. The directional response of a probe is required to follow a cosine law within an allowed tolerance. The tolerance on the angle for minimum intensity response is $\pm 5°$ about 90° for class 1 and $\pm 7°$ for class 2. This requirement is related to that for δ_{pI0} (see Section 6.3.1) and is, in fact, far less demanding than the latter. Directional response control is most important in excluding the intensity component tangential to a measurement surface generated by a strong extraneous source.

The Standard specifies a procedure for evaluating the operating range of a processor by determining the range of pink noise signal fed in common to both channels over which the indicated values of δ_{pIO} are invariant. When input signals become too small, the effect of noise described above will place a 'floor' on the residual intensity, thereby progressively reducing δ_{pIO} as the input signal is decreased. This problem may occur, for example, in measurements of flanking transmission in buildings in which the intensity level may be rather small. The main effect of electrical noise, however, is to increase random error.

The probe pressure and intensity response to a plane progressive wave field must lie within a stated tolerance of the nominal value. It is, in principle, possible to approximate such test conditions by making measurements in an anechoic chamber at ten or more diameters' distance from a compact source. However, readers are advised to determine precisely the variation of phase with radial distance from the source with a small single microphone before subjecting an intensity probe to this test: the fault might lie in the room and not in the probe! Alternatively, the various forms of gated or transient test techniques now available may be applied in a reflective room; however, the size of the room necessarily limits the available frequency resolution.

For use at frequencies below 400 Hz, a probe must also be tested in a standing wave field which imposes a maximum axial gradient of phase near to pressure minima and a maximum axial gradient of pressure halfway between pressure maxima and minima. This test applies in particular to sensors which have more than one entry port communicating with the sensing element; for example, the breather vent of a condenser microphone. The test may be made in a special standing wave tube large enough in cross-section to restrict the blockage due to the presence of the probe to less than 10% and equipped with gear to traverse it along the tube. Alternatively, a standing wave pattern may be traversed across a fixed probe by moving loudspeakers axially along the tube. It was not found possible to generate a reliable standing wave field in an anechoic chamber.[66] The indicated sound intensity level must lie within stated tolerances at all positions within the standing wave.

In addition to the usual pressure sensor calibration which must be applied to each channel in turn, a sound field of known intensity may be simulated by applying sound pressures of equal amplitude but different phase to the probe sensors. A residual intensity testing device which applies identical pressures to both sensors is not essentially a calibrator, but a means of evaluating the phase mismatch of the measurement system, as described above. A combined calibrator and residual intensity tester is shown in Fig. 6.26 in the following section.

Prior to the use of sound intensity measurement equipment, a minimum check comprising pressure sensor calibration and pressure-residual intensity evaluation *must* be made.

6.9   SYSTEM PERFORMANCE EVALUATION
AND FIELD CALIBRATION

6.9.1   p-p Systems

The simplest procedure for performing pressure calibration of the individual pressure transducers, while simultaneously evaluating the performance limit caused by phase mismatch, is to subject both transducers to the same sound pressure; the indicated particle velocity and intensity should then be zero. The phase mismatch may then either be measured directly, or inferred from the residual indicated values which are functions of the phase mismatch, as described in Section 6.3.1.

There are a number of possible techniques for generating and applying equal pressures to each transducer, as illustrated in Fig. 6.24. Great care has to be taken in implementing these techniques because small geometric details and imperfections can significantly influence the spatial distribution of phase, and minute vibrations of the mechanical structures employed can also adversely affect accuracy. The free field technique is most unreliable and is not

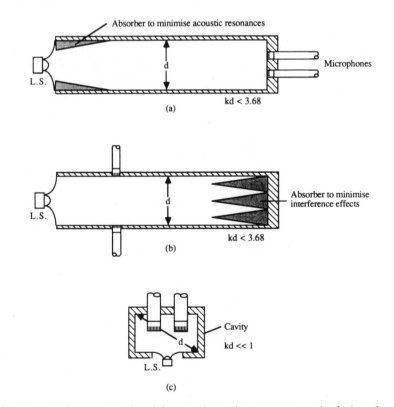

Fig. 6.24.   Various means of applying equal sound pressures to a pair of microphones.

recommended. Even though transducer reversal does, in principle, eliminate the influence of field non-uniformities, the physical problem of achieving geometrically precise reversal relegates this procedure to the rank of a performance check, and not a calibration. The other techniques all have upper frequency limits beyond which field uniformity is not sustained: this is not too serious since p-p phase mismatch effects are generally most severe at low frequencies. Configuration (a) is superior to configuration (b) because the microphones are at a pressure maximum at all frequencies, and hence errors in the estimate of the transfer function due to poor signal-to-noise ratio are minimised. It is also preferable to place an absorbent at the source end to minimise spectral irregularity due to acoustic resonance, as recommended for two-microphone impedance measurements in tubes.[67]

It is most important that the pressure equalisation vents of all condenser microphones, except the most recent intensity microphones which have very effective low pass vent filters,[56] should be subjected to the same pressure as the microphone diaphragm. As mentioned in Section 6.2.1, the fact that a vent and a diaphragm are exposed to different parts of a sound field can produce errors in fields where amplitude and phase vary rapidly with distance.

Because of difficulty in guaranteeing spatially uniform fields, it is always advisable to repeat this type of test with the transducer locations exchanged. In this way any small irregularities of phase may be detected and their effect eliminated by averaging the two results. Since $\delta_{pI} = 10 \lg |kd/(\phi_f \pm \phi_s)|$, evaluation of δ_{pI} in each case will allow the field phase difference to be separated from the measurement system phase mismatch; let the subscripts 1 and 2 denote the two tests:

$$\phi_s = \pm (kd/2)[10^{(-\delta_{pI(1)}/10}} - 10^{(-\delta_{pI(2)}/10)}] \tag{6.20a}$$

$$\phi_f = \pm (kd/2)[10^{(-\delta_{pI(1)}/10)} + 10^{(-\delta_{pI(2)}/10)}] \tag{6.20b}$$

The signs in eqns (6.20) are determined by the sense of I measured in each case. The corresponding pressure-residual intensity index is given by

$$\delta_{pI0} = 10 \lg |kd/\phi_s|$$
$$= [3 - 10\lg[\pm 10^{(-\delta_{pI(1)}/10)} \mp 10^{(-\delta_{pI(2)}/10)}] \, dB \tag{6.21}$$

In practice, instrument output resolution in dB may be inferior to the resolution of linear quantities, and it may be better to evaluate ϕ_s from cross-spectral data, or to use the linear relationships corresponding to eqns (6.20) and (6.21).

As shown by eqn (6.3), the normalised bias error in measured intensity is related to the difference L_ϕ between the pressure-residual intensity index δ_{pI0}, which is characteristic of the measurement instrument, and the pressure-intensity index δ_{pI} which is a function of the field and the probe orientation:

$$e_\phi(I) = [antilog(L_\phi/10) - 1]^{-1} \tag{6.22}$$

It was suggested earlier that L_ϕ should be known as the Phase Error Index,

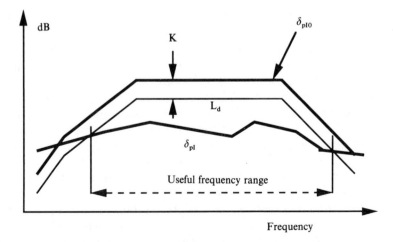

Fig. 6.25.　Definition of the Dynamic Capability Index L_d.

since it indicates the severity of the effect of instrument phase mismatch on the accuracy of any one measurement.

If a limit is placed on the maximum acceptable normalised bias error, a criterion for acceptable instrument performance may be defined in terms of L_ϕ. For example, if we wish to limit $e_\phi(I)$ to no more than ± 0.25 (corresponding to ± 1 dB), the corresponding minimum value of L_ϕ is approximately 7 dB. The difference between L_ϕ and L_ϕ (min) is termed Dynamic Capability L_d. Figure 6.25 shows how L_d may be used to specify the usable frequency range of an instrument in terms of a maximum acceptable bias error due to phase mismatch.

It is possible to simulate a sound field of known intensity in a cavity by arranging suitable acoustic elements between the microphone positions, so that a known phase difference is produced.[68] Thus, particle velocity and intensity calibration may be performed. A currently available calibrator is shown, together with its characteristics, in Fig. 6.26.

6.9.2　p-u Systems

The pressure transducer may be calibrated in the normal way. The particle velocity transducers may be calibrated in a spherical progressive sound field against a pressure measurement, and the application of the relationship $u = p/\rho_0 c$. Phase mismatch may be evaluated at low frequencies in a standing wave tube, as described in Section 6.3.2: particle velocity calibration may also be achieved by this technique, under various conditions of phase gradient. It is extremely difficult to measure phase mismatch at high frequencies, although it may be inferred from measurements in a spherical progressive sound field once the pressure and particle velocity sensitivities have been established. At best this is a check, and not an accurate means of calibration. The variation of

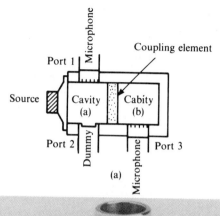

(a)

(b)

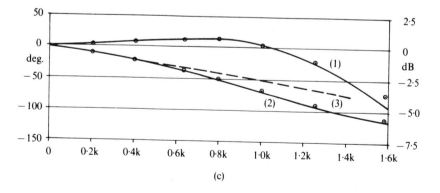

(c)

Fig. 6.26. Acoustic calibrator for p-p intensity measurement systems; (a) schematic; (b) photograph; (c) performance: ((1) magnitude ratio; (2) phase difference; (3) free field phase) (courtesy Brüel & Kjaer, Denmark).

sensitivity of a p-u probe with temperature should be specified by the manufacturer.

The relationship between δ_{pI} and the error due to phase mismatch is different from that of a p-p probe because it depends only on field characteristics and not on probe geometry. According to eqn (6.6)

$$e_\phi(I) \approx \phi_s \tan(\phi_f) \tag{6.23}$$

This relationship is plotted in logarithmic form in Fig. 6.16.

CHAPTER SEVEN

Errors in sound intensity measurement

7.1 INTRODUCTION

Measurements of sound intensity are subject to errors associated with the following factors: approximations made in the assumed relations between the directly transduced quantities and the intensity (inherent errors); imperfections of the probe as a transducer; imperfections of the signal processor in its function of converting the acquired analogue signals into the quantities required to compute the intensity; errors of calibration; variations of transducer sensitivity from the calibration value caused by environmental conditions; 'noise' produced either by non-acoustic disturbances, such as turbulent airflow, or by the instrument itself; and errors associated with the time-averaging/integration process used in spectral estimation procedures. The process of converting estimates of sound intensity into estimates of source sound power is subject to additional sources of error which will be treated in Section 9.3.

Some of these sources of error have been mentioned in earlier chapters. Errors which are inherent in the techniques of sound intensity measurement have been dealt with in Section 5.6; probe performance is the subject of Sections 6.3–6.5; instrument performance and field calibration are treated in Sections 6.6–6.9. This chapter deals first with the systematic error produced by phase mismatch between the transducers and the associated signal conditioning channels which is related to the form of the acoustic field in which a measurement is being made: methods of correcting for this form of error follow. The chapter continues with an analysis of random errors associated with spectral estimation as functions of the signal processing procedures employed and the form of the acoustic field being investigated.

7.2 SYSTEMATIC ERROR DUE TO PHASE MISMATCH

Microphones respond not only to acoustic fields but also to non-acoustic pressure fluctuations in unsteady fluid flow, as described in Section 6.4. These can severely compromise the accuracy of sound intensity measurements, especially when the non-acoustic pressures are partially correlated, as they would be if the p-p sensors are close together in terms of the spatial correlation length of the unsteady flow. It is assumed in this chapter that the pressures sensed are entirely acoustic. The relationship between the mean square pressure at a point in a single frequency sound field and the estimated intensity component in the direction of the axis of a p-p probe is given by the approximation to eqn (4.54) expressed by eqn (5.7), on condition that $P_1 P_2/2$ is a good approximation to the mean square pressure $\overline{p^2}$ midway between the acoustic centres of the pressure sensors and that the field phase difference $\phi_f = \phi_1 - \phi_2$ (radians) between these points is much less than unity. (Note: these conditions require the finite difference approximation errors in the estimates of $\overline{p^2}$ and ϕ to be sufficiently small (see Section 5.6).) The positive intensity direction is from the acoustic centre of sensor 1 to that of sensor 2. We may therefore write

$$I = (\overline{p^2}/\rho_0 c)\phi_f/kd \qquad (7.1)$$

in which I is the component of the intensity vector directed along the probe axis. At any frequency, the effect of a phase mismatch ϕ_s between the two channels of a transducer-signal conditioning chain is to bias this estimate to

$$I_e = (\overline{p^2}/\rho_0 c)(\phi_f + \phi_s)/kd \qquad (7.2)$$

in which it is assumed that the estimate of $\overline{p^2}$ is unbiased. If ϕ_s is of opposite sign to ϕ_f, the intensity will be underestimated; and if it is also greater than ϕ_f, the estimated and true intensity components will be oppositely directed. The normalised bias error is

$$e_\phi(I) = (I_e - I)/I = \phi_s/\phi_f \qquad (7.3)$$

If a p-p probe is placed in a sound field in which the pressure sensors are exposed to identical pressures (i.e. $\phi_f = 0$), the indicated 'residual' intensity is given by eqn (7.2) as

$$I_{res} = (\overline{p^2}/\rho_0 c)_{res}(\phi_s/kd) \qquad (7.4)$$

in which the subscript 'res' on the pressure term serves only to indicate the particular value which exists in this special field condition. (Note: ϕ_s, and therefore I_{res}, may be positive or negative.) Combining eqns (7.1), (7.2) and (7.4) yields

$$I_e - I = I_{res}[(\overline{p^2}/\rho_0 c)/(\overline{p^2}/\rho_0 c)_{res}] \qquad (7.5)$$

We will denote this quantity by D. When the characteristic impedance of the fluid is the same in both test and calibration conditions,

$$D = I_{res}[\overline{p^2}/p_{res}^2] \qquad (7.6)$$

The corresponding normalised bias error is

$$e_\phi(I) = (I_e - I)/I = D/(I_e - D) \qquad (7.7)$$

This error may also be expressed in terms of difference between the pressure-residual intensity index δ_{pI0} (formerly termed the 'residual pressure-intensity index') and the observed pressure-intensity index δ_{pI} at the field point (eqn (6.3)). (Note: the observed pressure-intensity index is biased if the intensity estimate is biased.) By definition

$$\overline{p^2}/\rho_0 c = |I_e| 10^{\delta_{pI}/10} \tag{7.8}$$

and

$$(\overline{p^2}/\rho_0 c)_{res} = |I_{res}| 10^{\delta_{pI0}/10} \tag{7.9}$$

Hence the term D may be written

$$D = \pm I_e 10^{-(\delta_{pI0} - \delta_{pI})/10} \tag{7.10}$$

The quantity $\delta_{pI0} - \delta_{pI}$ has been termed the phase error index and denoted by L_ϕ in Section 6.3.1. Since the signs (directions) of intensity are suppressed in the definition of the pressure-intensity indices, the positive sign is chosen if both I_e and I_{res} have the same sign (i.e. direction by reference to the calibration data). The ratio of estimated to true intensity is given by

$$I_e/I = [1 \pm 10^{L_\phi/10}]^{-1} \tag{7.11a}$$

where the negative sign is chosen if I_e and I_{res} have the same sign. The normalised bias error is given by

$$e_\phi(I) = [\pm 10^{L_\phi/10} - 1]^{-1} \tag{7.11b}$$

where the positive sign is used when I_e and I_{res} have the same sign.

These equations are valid provided that the bias error is not so great as to make I_e/I negative. The same equations give the phase mismatch bias error in an estimate of the spectral density $I(\omega)$ of the active intensity of a time-stationary aperiodic field by replacing the mean square pressure by its auto-spectral density and the pressure-intensity indices by their spectral equivalents.

The expressions above apply to single frequency quantities and to spectral densities. In practice, intensity estimates are usually made in finite frequency bands—most frequently 1/3 octave bands. The question arises as to the applicability of these relations to frequency band measurements. It is clear that in cases of aperiodic, time-stationary sound fields eqn (7.5) may be expressed in terms of a summation of spectral energies, or an integral of spectral densities, over the band concerned. If the phase mismatch ϕ_s varies slowly over the band, which is generally the case in practice, the term $I_{res}/(\overline{p^2}/\rho_0 c)_{res}$ may be excluded from the operation of summation or integration, and a frequency-average value substituted. Hence the frequency band equivalent of eqn (7.5) is

$$I_e(\Delta\omega) - I(\Delta\omega) = \langle (I_{res}/(\overline{p^2}/\rho_0 c)_{res}) \rangle \int_{\Delta\omega} (\overline{p^2}/\rho_0 c) \, d\omega \tag{7.12}$$

in which $\langle \, \rangle$ denotes 'average in the band'.

An expression for the normalised systematic error equivalent to eqn (7.7) may be written with D equal to the term on the r.h.s. of eqn (7.12).

Similarly, a frequency band equation equivalent to eqn (6.3) may be written in terms of the pressure-intensity index measured in the band. The frequency-band pressure-intensity index is given by

$$\langle \delta_{pI} \rangle = 10 \lg \int_{\Delta\omega} (\overline{p^2}/\rho_0 c) \, d\omega - 10 \lg \int_{\Delta\omega} I_e \, d\omega \qquad (7.13)$$

Hence, the integral in eqn (7.12) may be written as

$$\int_{\Delta\omega} \overline{p^2} \, d\omega = |I_e(\Delta\omega)| 10^{\langle \delta_{pI} \rangle / 10}$$

and D may be expressed as

$$D = \pm I_e(\Delta\omega) 10^{(\langle \delta_{pI} \rangle - \langle \delta_{pI0} \rangle)/10} \qquad (7.14)$$

in which $\langle \delta_{pI0} \rangle$ corresponds to the pressure-residual intensity index measured using a uniform pressure spectrum within the band. The normalised bias error is given in logarithmic form by

$$e_\phi(I) = [\pm 10^{\langle \delta_{pI0} \rangle - \langle \delta_{pI} \rangle} - 1]^{-1} \qquad (7.15)$$

As before, the positive sign is used if I_e and I_{res} have the same sign.

The plots of normalised systematic error and its logarithmic equivalent presented in Fig. 6.11 as a function of L_ϕ for single frequencies also applies to frequency bands where L_ϕ is taken as the difference between $\langle \delta_{pI0} \rangle$ and $\langle \delta_{pI} \rangle$.

For a given value of L_ϕ, the magnitude of the normalised error due to phase mismatch which produces an underestimate is less than that which produces an overestimate. Equations (7.11) and (7.15) may, in principle, be used to correct an estimate of intensity. However, the accuracy of the correction is dependent upon the accuracy with which $\langle \delta_{pI} \rangle$ and $\langle \delta_{pI0} \rangle$ can be determined.[69] In very difficult field conditions, the difference between $\langle \delta_{pI} \rangle$ and $\langle \delta_{pI0} \rangle$ will be small, and the uncertainty of determination of $\langle \delta_{pI} \rangle$ associated with random error may be of the same order as this difference, in which case confidence in the correction will be low. As will be shown in the following sections on random errors, phase mismatch which changes the apparent sign of the phase difference (and intensity) of some frequency components within an analysis band greatly increases the normalised random error of the estimate of intensity in that band: hence, even if the phase mismatch error is well known, any associated correction will be applied to a less reliable estimate. There exists, therefore, a kind of 'coupling' between the effects of systematic and random error.

7.2.1 Methods of correcting for phase mismatch

The phase mismatch between the two channels of a p-p instrument can be determined by means of an intensity calibrator from eqn (7.4) or its equivalent expressed in terms of the pressure-residual intensity index δ_{pI0} as

$$\phi_s = \pm (kd)10^{-\delta_{pI0}/10} \qquad (7.16)$$

The sign of ϕ_s is given by the sign of the indicated residual intensity. The estimated intensity may therefore be corrected as indicated by eqns (7.11) and (7.15).

In terms of the transfer function H_{12} between the signals measured when the sensors are exposed to identical pressures, the equivalent correction process is expressed as

$$G'_{12} = G_{12}/(H_{12}|H_1|^2) \qquad (7.17)$$

where G' is the corrected estimate of the pressure cross-spectral density, and H_1 is the pressure sensitivity of channel 1 which is treated as the 'input' channel.[70] This technique has been elaborated[71] to include transducer exchange in order to eliminate the effects of non-ideal test fields. This correction may be implemented in software used for postprocessing the output from an FFT analysis. The estimates of the transfer function in eqn (7.17) are themselves subject to random error, but the signals from pressure transducers placed in a small cavity are likely to be free of ambient noise and fully coherent, with a rather uniform spectrum; random errors will therefore be small. However, care should be taken to minimise ambient noise and mechanical vibration of the cavity structure.

A simple but effective means for minimising bias error due to phase mismatch is to employ probe reversal in a similar manner to that described in Section 6.9.1 on probe calibration. The arithmetic average of the two intensities indicated by an instrument with the probe first placed at a measurement point and then reversed *about its effective measurement point* is, in principle, unbiased, irrespective of the signal or analysis bandwidth. Practical difficulties are created by the fact that intensity probes are usually not geometrically symmetric about their measurement point, that they can often not be reversed when the measurement point is near a solid surface and that precise reversal is difficult unless a mechanical support is used. The field must of course remain unchanged for the successive measurements, and scattering from any objects which change position during reversal can lead to serious errors.

An alternative technique for phase mismatch correction which is also based upon probe (or microphone) reversal is described by Jacobsen[72] and Ren and Jacobsen.[73] The probe is placed in the near field of a large loudspeaker, preferably with the probe axis coincident with the cone axis, and the difference between indicated intensities yields the pressure-residual intensity index (see eqn (6.21)). Uncertainty about the invariance of the field conditions during physical reversal preclude the use of this technique for making precision corrections to systems which are already very well matched.

It should be emphasised that the random error of determination of the phase mismatch can be large, especially when the phase mismatch is very small: electrical noise and extraneous acoustic noise and vibration can randomly vary the indicated sign of the residual intensity, which therefore demands a very large BT product to reduce the random error to insignificance. Every effort should therefore be made to minimise these adverse influences. There is a practical minimum to which phase mismatch can be reduced by correction because of the influence of random error and it is vital that the repeatability of phase correction which is applied to any individual instrument is firmly established and that the correction is confirmed immediately prior to its use.

7.3   RANDOM ERROR ASSOCIATED WITH SIGNAL PROCESSING

Even if the systematic (bias) errors of an intensity estimate can be accurately quantified, there remains inherent uncertainty about the difference between an estimated value of any derived quantity (output) indicated by any signal processing device and the true value of that quantity. This difference is termed 'random error'. Random error may be generated by inherent limitations or deficiencies of the instrumentation, such as internally generated random noise; by the action of non-acoustic pressures on the sensors, such as those of turbulent flow; or by limitations of the processing technique used to produce the estimate—in particular, finite processing time. This section deals with the last source of error, which is dependent not only upon the nature of the signals and signal processing procedures but also upon the form of the sound field under investigation.

By its very nature, random error can only be quantified in statistical terms. The most complete statement of uncertainty takes the form of a confidence estimate; for example, there is $x\%$ probability that the true value lies in a range having limits $e(1 + y)$ and $e(1 - z)$, where e is the estimated value. These values are the $x\%$ confidence limits, and $e(y + z)$ is the $x\%$ confidence interval. Ideally, the uncertainty of an estimate should be expressed in the continuous curve of y and z as a function of x. However, in many cases of practical interest, the probability distributions of estimates are found to approximate closely to one of the ideal distributions such as Gaussian (normal), Poisson, gamma, etc. In such cases, it is only necessary to estimate the mean and standard deviation (or variance) of the estimate, from which any confidence interval can be determined.

In statistical analyses of the uncertainty of spectral estimates derived from signals generated by transducers operating on linear acoustic processes, it is commonly assumed that the estimates are normally distributed. This assumption is based upon the fact that an estimate of a spectral quantity is obtained by averaging a number of individual estimates which are, ideally, *statistically independent*; in this case, the central limit theorem states that the estimate of the *mean* will be normally distributed and the variance of the estimate of the mean will be inversely proportional to the number of individual independent estimates, irrespective of the distribution of the estimates themselves. It is clear that estimates based upon a series or sequence of spectral analyses of a signal, or signals, over discrete, non-overlapping intervals of time are independent only if the correlation function of the signal(s) decay(s) to insignificance in a time which is shorter than the length of the discrete intervals: otherwise, successive estimates are partly based upon statistically related data.

The decay rate of the autocorrelation function of a time-stationary, band-limited, aperiodic signal is inversely proportional to its frequency bandwidth. In cases of analysis by digital or analogue band pass filters, the decay rate is determined by either the bandwidth of the signal or of the filter, whichever is the narrower (only for signals of uniform spectral density are they the same). When signal analysis is performed by Discrete Fourier Transform (DFT) procedures, and the results are presented in frequency bands containing a number of spectral lines, the relevant autocorrelation decay rate is determined by the width of the frequency band of presentation, not by the spectral resolution of the analysis, provided that the signal does not actually have a narrower bandwidth. The decay of the cross-correlation function of two acoustic pressure signals from an intensity probe depends not only on their bandwidth but also upon the presence of any reflections of sound from objects or enclosure boundaries; for example the multiple reflection process in a room. If the sequence of reflections arrives with delay times in excess of $1/B$, where B is the analysis bandwidth, the variance of the estimates of spectral quantities will not vary as $(BT)^{-1}$, where T is the averaging time.

In the analyses which follow it is implicitly assumed that the signals are random and 'white' (uniform spectral density) and have normal amplitude distributions or joint amplitude distributions. The reason is one of analytical expediency and convenience; the reader is warned that many real signals do not possess these ideal qualities.

7.3.1 Analysis of random error in intensity estimates

This section analyses random error associated with estimates of intensity based upon signals generated by an intensity probe located at a fixed point in a time-stationary sound field: it does not deal with uncertainty associated with moving probe (scanned) measurements, or with non-stationary signals. The effect of scanning is discussed in Section 9.3.

The spectral density of active intensity is proportional to the real component C_{pu} of the cross-spectral density of signals from a p-u probe and to the imaginary component Q_{12} of the signals from a p-p probe. On the basis of the assumption that the auto- and cross-correlation functions of the signals decay quickly in comparison with the total integration time T of the spectral estimation process, Jacobsen[74] derives two general expressions for the random error of an intensity estimate. In terms of the complex cross-spectrum $C_{pu_r} + iQ_{pu_r}$ of sound pressure and the component of particle velocity in direction **r**, the variance of an estimate of the spectral density of intensity at frequency ω is given by

$$\sigma^2\{\hat{I}\} = \int_0^\infty (S_{pp}(\omega)S_{u_ru_r}(\omega) + C_{pu_r}^2(\omega) - Q_{pu_r}^2(\omega))\,d\omega \qquad (7.18)$$

in which S_{pp} and $S_{u_ru_r}$ are auto-spectra and T is the integration time. The corresponding expression for the normalised random error of an estimate of intensity made in a finite frequency band $\Delta\omega$ is

$$\varepsilon\{\hat{I}_r(\omega_0, \Delta\omega)\} \approx \frac{\left(\int_{\omega_a}^{\omega_b} [S_{pp}(\omega)S_{u_ru_r}(\omega) + C_{pu_r}^2(\omega) - Q_{pu_r}^2(\omega)]\,d\omega\pi/T\right)^{1/2}}{\left|\int_{\omega_a}^{\omega_b} C_{pu_r}(\omega)\,d\omega\right|} \qquad (7.19)$$

which may also be written as

$$\varepsilon\{\hat{I}_r(\omega_0, \Delta\omega)\} \approx \frac{\left(\int_{\omega_a}^{\omega_b} [C_{pu_r}^2(\omega)(1 + 1/\gamma_{pu_r}^2(\omega)) + Q_{pu_r}^2(\omega)(1/\gamma_{pu_r}^2(\omega) - 1)]\,d\omega\pi/T\right)^{1/2}}{\left|\int_{\omega_a}^{\omega_b} C_{pu_r}(\omega)\,d\omega\right|}$$

$$(7.20)$$

in which γ^2 is the coherence between p and u_r (see Fig. 7.1).[75]

The corresponding frequency band expressions in terms of the auto- and cross-spectra of signals from a p-p probe are

$$\varepsilon\{\hat{I}_r(\omega_0, \Delta\omega)\} = \frac{\left(\int_{\omega_a}^{\omega_b} [(S_{11}(\omega)S_{22}(\omega) - C_{12}^2(\omega) + Q_{12}^2(\omega))/\omega^2]\,d\omega\pi/T\right)^{1/2}}{\left|\int_{\omega_a}^{\omega_b} (Q_{12}(\omega)/\omega)\,d\omega\right|} \qquad (7.21)$$

and

$$\varepsilon\{\hat{I}_r(\omega_0, \Delta\omega)\} \approx \frac{\left(\int_{\omega_a}^{\omega_b} [(Q_{12}^2(\omega)(1 + 1/\gamma_{12}^2(\omega)) + C_{12}^2(\omega)(1/\gamma_{12}^2(\omega) - 1))/\omega^2]\,d\omega\pi/T\right)^{1/2}}{\left|\int_{\omega_a}^{\omega_b} (Q_{12}(\omega)/\omega)\,d\omega\right|}$$

$$(7.22)$$

In the very unlikely case that all spectral quantities are independent of frequency, eqn (7.22) is equivalent to that published by Pascal,[76] and very similar to that of Seybert.[77]

At first sight, eqns (7.20) and (7.22), of which the integrands have similar form, would appear to give different estimates in 'nearly' diffuse fields in which $\gamma_{pu_r}^2$ is close to zero whereas γ_{12}^2 remains close to unity because the non-dimensional distance kd between the p-p probe microphones must be much

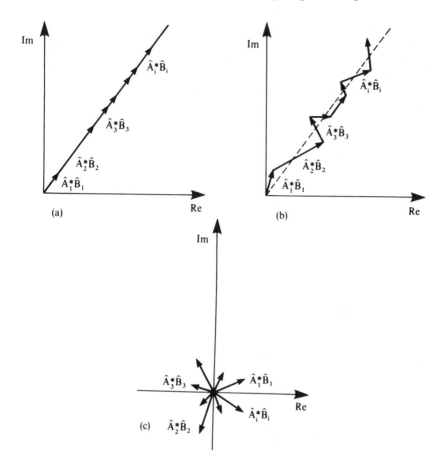

Fig. 7.1. Illustration of the significance of the coherence function: (a) phase of each estimate the same; (b) some fluctuation in the phase of successive estimates; (c) random fluctuation in the phase of successive estimates.[75]

less than unity. However, in such fields $C_{\text{pu}_r}, Q_{\text{pu}_r}$ and Q_{12} are all small, whereas C_{12} remains large.

Given that $Q_{12} = -\omega \rho_0 dI$ (eqn (5.19)) and $S_{11}S_{22} \approx S_{\text{pp}}^2$, the term in the numerator of eqn (7.21) may be approximated as

$$S_{11}S_{22} - C_{12}^2 + Q_{12}^2 = Q_{12}^2[((1 - \gamma_{12}^2)/(kd)^2)((S_{\text{pp}}/\rho_0 c)^2/I^2) + 2]$$

$$(7.23)$$

which form indicates that the pressure-intensity index δ_{pI} is only of consequence when $\gamma_{12}^2 < 1$. It also suggests that δ_{pI} is only significant if it satisfies the condition $10^{\delta_{\text{pI}}/5} > 2(kd)^2/(1 - \gamma^2)$.

Unfortunately, it is not generally practicable to employ these equations to make precise estimates of random errors, for the following reasons. First, estimates of the coherence function are usually subject to serious bias error due to inadequately fine frequency resolution, especially in reverberant conditions[78] when the sound field has an 'integration time' $T_{60}/6\cdot9$; second, unless the analysis bandwidth B is much less than the individual resonance bandwidth, which is equal to $2\cdot2/T_{60}$, estimates of spectral quantities which enter the equations are themselves subject to random errors which are functions of the coherence function; and third, narrow band coherence estimates are not available from instruments based upon analogue or digital filters—the so-called 'frequency band coherence function' obtainable from some of these instruments may not, except under very special conditions, be validly used for error analysis.

The following important conclusions can be drawn from the above equations. Because Q_{12}/ω appears as the integrand in the denominator of eqns (7.21) and (7.22), whereas its square appears in the integrand of the numerator, any variation of the intensity over the frequency band, especially in direction (sign), will increase the ratio of numerator to denominator; hence, the normalised error will increase. A similar observation applies to C_{pu_r} in eqns (7.19) and (7.20). (Of course, a large uncertainty in a small quantity may be unimportant in practice.) If the coherence between pressure signals is unity, and the intensity does not vary significantly within the analysis band, the normalised random error in the intensity estimate is approximately equal to $(BT)^{-1/2}$, where B is the analysis (not resolution) bandwidth and T is the effective averaging time which is less than the measurement time unless overlapped real-time processing is performed. In fact, the p-p coherence is generally close to unity in all sound fields, because of the closeness of the microphones.[79]

Equation (7.20) indicates that the random error of an estimate made in a highly reactive field, such as the near field of a source or in a strong standing wave field, is very sensitive to the presence of a 'diffuse' component which reduces $\gamma^2_{pu_r}$ below unity, because $Q_{pu_r}/C_{pu_r} \gg 1$. The pressure-intensity index will be large in near fields and reverberant fields, and often in strong standing wave fields, but the influence of δ_{pI} will only be significant if γ^2_{12} is less than unity as will be the case at high frequencies in diffuse fields, and also where the acoustic pressure signals are contaminated by signals produced by unsteady flow, electrical noise or vibration of probe or cables.

CHAPTER EIGHT

Classification and characterisation of sound sources and fields

8.1 INTRODUCTION

This chapter discusses source classification in terms of effects, as opposed to generic forms of mechanism described in Section 3.6.1. A qualitative classification of the various regions of sound fields generated by sources in terms of sound intensity distributions is followed by a section describing how the results of two-microphone (p-p) measurements may be used to classify and characterise sound fields in quantitative terms, with particular emphasis on the significance and application of 'sound field indicators'.

Sound fields differ widely in their temporal and spatial characteristics: these are determined by the nature, form and location of the sound-generating mechanisms, the form and properties of the boundaries of the fluid medium, and any spatial variation of sound speed within the medium. A complete description of any but the simplest sound fields is unattainable. However, for the purpose of discussing their energetic features it is useful to classify forms and regions of sound fields according to a set of broad qualitative descriptions. As a prelude to addressing this matter, the chapter opens with a question— 'What is a *source*?' This question may seem unnecessarily enigmatic, but it stems from the fact that there are no universal criteria for separating a complex sound generator into discrete, individual sources: indeed, sources may be differently classified according to which physical effect of their actions is considered. It will be shown that the question of the degree of correlation between the actions of simultaneously operating sound generation mechanisms is of vital concern to the acoustician faced with the problem of taking action to reduce their effects. The chapter continues with qualitative descriptions of the various components and regions of sound fields generated by sources in free field and in the presence of reflective boundaries.

For the purposes of identifying sources, interpreting sound intensity measurements and selecting the most effective noise control measures, it would be very useful to be able to characterise sound fields on the basis of experimental data. Conventional instrumentation using a single pressure transducer can yield measures of the spatial average value and spatial distribution of mean square pressure and its associated sound pressure level. Also, by means of a 'walk-away' test, some indication may be obtained of the extents of the direct and reverberant fields of a source. Two-microphone (p-p) measurements may yield a much more comprehensive characterisation of a sound field because pressure gradient data, and the associated particle velocity data, are also acquired; the relationship between sound pressure and particle velocity is highly indicative of the nature of the field. In addition, direct evaluation may be made of the distributions of potential and kinetic energy densities. (Such determinations are not new; for example, Cook[80] constructed an energy density meter in 1974 for investigating reverberant sound fields and Riedlinger[81] developed an energy density measuring microphone.)

The final section of this chapter describes the various sound field indicators which have been introduced for the purpose of sound field characterisation. Additional indicators are introduced in Chapter 9 in connection with the application of sound intensity to the determination of source sound power. Much of the material in this section is drawn from the publications of Finn Jacobsen who has made a major contribution to recent developments in this area of application of p-p measurements.

8.2 WHAT IS A SOURCE?

Generators of sound are infinitely diverse in their physical forms and characteristics. In spite of this great diversity, each may be placed in one or more of the three categories introduced in Section 3.6.1. Although classification is useful for the purposes of idealised mathematical representation and analysis, it is not immediately obvious that such idealisation has any material bearing on the practice of sound intensity measurement. However, theoretical analyses of elementary source fields can be of considerable assistance in elucidating the causes of the (often puzzling) results of experimental measurement.

Before its realisation, the sound intensity meter was seen as a kind of acousticians' 'philosophers' stone'; just wave it around a source and you will know exactly where the noise comes from—and how to suppress it! Experience gathered over the past fifteen years has firmly disabused acousticians of that naïve concept, and also of the concept of 'The Source'.

Most sources encountered in engineering practice are extended in space, exhibit a mixture of idealised source characteristics, contain components which perform mutually correlated actions, and usually operate in the sound fields generated by other sources. 'The Source' of the troublesome agent, sound pressure at some point in space, may not correspond at all closely to 'The Source' as defined in terms of the generation of sound power. This is perfectly evident within enclosures such as vehicle passenger compartments at low audio-frequencies.

From a mathematical point of view, there is a certain ambiguity about the identification of 'The Source' of any given sound field. The acoustic pressure at a point in space is given by an integral, over *any* closed surface surrounding that point, of the products of the pressures and normal accelerations of the fluid on that surface with appropriate transfer (Green's) functions—the Kirchhoff–Helmholtz Integral.[33] Hence, any such surface which has an appropriate distribution of these two quantities could be considered to be a 'source' surface. Of course, most engineers would dismiss this view as obfuscation, because they can 'see' the source; for example, in the form of a vibrating solid surface, or a turbulent jet efflux. However, one valuable corollary of the mathematical non-uniqueness of sources is that, in attempting to identify and quantify physical sources, one needs criteria for positive identification.

In this respect, it should be realised that there is no simple relationship between 'sources' of sound pressure and 'sources' of sound intensity, because the latter involves the product of two field variables (or, alternatively, the product of the space and time derivatives of the scalar velocity potential). Hence, only under certain conditions will relationships between the measured quantity (intensity), and the troublesome quantity (pressure), be predictable and exploitable for the purposes of noise control; and then only in terms of the total sound power of a source, rather than any local intensity. These are conditions in which the injection of a certain amount of sound power into a volume of fluid generates mean square pressure in some predictable proportion. The principal physical feature of sound fields which inhibits any universal relationship is that of interference. The sound power radiated by a distributed source depends upon the 'strength' and spatial distribution of its fundamental mechanism (e.g. surface acceleration, fluctuating force, fluctuating mass introduction), and also on the acoustic load, or radiation impedance, experienced by the mechanism. By contrast, the field pressure at points removed from the source region is determined not only by the source mechanism, but by local wave interference behaviour which has no influence on the source power. For example, the sound power generated by a harmonic point monopole (say a small loudspeaker at low frequency) can be varied over a wide range by bringing into its vicinity a large, plane, rigid surface. However, the pressure amplitude at any point on that plane surface is twice that produced at the same point by the isolated monopole, irrespective of the distance between the source and surface.

The situations in which source sound power and sound pressure bear simple relationships to each other are as follows: (i) in the geometric far field of any source in free field; (ii) in a diffuse reverberant field; (iii) in an anechoically terminated, uniform duct below its lowest cut-off frequency. In the first case, a complete description of the relationship demands knowledge of the source directivity. Modification of a source to reduce its power is likely to result in a change in its directivity, so that sound pressure at any given point may not fall in concert; however, the spatial average mean square pressure over an enveloping sphere must, of course, do so. Measures which reduce the power of a broad band source operating in a reverberant enclosure of dimensions large compared with the longest significant acoustic wavelength will bring about a concomitant change in the spatially almost uniform mean square pressure. This will not, however, necessarily be the case for harmonic, or narrow band, noise sources, because the interference pattern everywhere in the enclosure depends upon the free field source directivity, as well as on the enclosure geometry and boundary conditions.

The distinction between 'power sources' and 'pressure sources' may be further elucidated by appeal to the principle of reciprocity. In any given, time-invariant, linear fluid system, the transfer function between the volume velocity of a point monopole source and the pressure at any observation point in the acoustic field is invariant with respect to exchange of the source and observation points. Since the motion of a small element of a vibrating surface of a body may be represented by an elemental volumetric source, it is clear that the relative influence of the vibration of each element of the surface on the pressure at some field point may be determined by placing a small omni-directional source at the field point of interest, and measuring the transfer function to the pressure on the *rigid* surface of the body.[33] The integral over the surface of the body of the product of this transfer function and the actual normal surface velocity of the vibrating body yields the field pressure at the field point concerned. Since the transfer function varies with the location of this point, the 'pressure source' distribution varies with the field point considered. By contrast, the 'power source' distribution, although dependent upon the shape of the body, its vibration distribution, and the acoustic characteristics of the whole environment in which the body is situated, is not observation-point dependent. Only in special cases such as the diffuse reverberant field, is the transfer function largely independent of the location of the observation point, in which case the effects of modifying source power are reproduced in the pressure. The practical implication of the distinction discussed above is that any change made to the vibration distribution of a radiating surface will normally have quite different effects on the pressure observed at any field point, and on the surface intensity distribution, and hence, on radiated power.

The purpose of the aforegoing *caveat* is to serve as a warning to the unwary, not as deterrent to the uninitiated. Sound intensity measurement is a very useful weapon in the armoury of the acoustical engineer, but it does not necessarily provide the answer to the question 'What should I do to reduce the noise?'

8.3   CHARACTERISTICS OF SOUND FIELDS NEAR SOURCES

Most intensity measurements are made fairly close to the regions of generation of sound, whether the sources be industrial machines, workpieces, boundaries of vehicle cabins, building partitions, musical instruments, jet mixing regions, or one of the other thousand-and-one forms encountered by the noise control engineer. The acoustic characteristics of the spaces into which such sources radiate are extremely diverse, ranging from highly reverberant spaces containing many sources, such as factories, to largely non-reflecting outdoor environments.

At any point in the vicinity of a (primary) source under consideration, the total acoustic field may be decomposed into three components: (i) the field directly radiated by the primary source (i.e. that which would exist in free field); (ii) the corresponding fields radiated directly from other sources present (which may be parts of the mechanical system incorporating the primary source but which lie outside the chosen measurement surface); (iii) the field produced by reflection, scattering and diffraction of these directly radiated fields by bodies and bounding surfaces. These three components contribute to the total mean square pressure at any point in a manner depending upon their degree of mutual correlation. The resulting intensity distribution is, in addition, dependent on the distribution and correlation of particle velocity vectors.

The direct field of a source may itself be divided into three regions: (i) the hydrodynamic near field, in which corresponding frequency components of particle velocity and pressure are nearly in quadrature, and in which $|u| \gg |p|/\rho_0 c$; (ii) the geometric near field, in which the source subtends a large angle at an observer point, in which p does not vary inversely with distance from the centre of the source, and in which neither the particle velocity nor intensity vectors are necessarily directed radially from the source centre, although the phase angle between pressure and particle velocity may be quite small; (iii) the far field, in which the source subtends a small angle, I and u are radially directed, and p and u are in phase. This decomposition is valid provided that, in this context, a 'source' comprises the whole extent of any fully, or partially, correlated sound generation system; it is not amenable to an arbitrary selection of 'the source' as defined by any convenient choice of intensity measurement surface. Circulatory patterns of active sound intensity can exist under free field conditions both in the hydrodynamic and geometric near fields, but not in the true far field, where particle trajectories are essentially directed along radial lines.

Most intensity fields encountered in practice are too complicated to be precisely modelled mathematically, and it is therefore necessary to consider idealised field models in order to appreciate the combined effects of these various field components. The effects of coherent extraneous sources and reflections in modifying the intensity distribution and source sound power output will be discussed in Section 9.2.

8.3.1 Idealised field models

(a) Hydrodynamic near fields of vibrating surfaces

Sound fields generated by vibrating solid surfaces have been analysed in Sections 3.8, 4.8.6 and 4.8.7. In the region of the hydrodynamic near field, single frequency particle trajectories are elliptical; the particle velocity amplitudes are much greater than $|p|/\rho_0 c$; the pressure gradients, and therefore reactive intensity components, are strong relative to active components; and the active intensity exhibits circulatory patterns, with the vibrating structure forming part of the power flow circuit. The pressure-intensity index in such fields is generally high and, in principle, unlimited; the finite separation errors produced by p-p probes can be large; and, because of power circulation, any attempt to rank order regions by near field measurement is not a reliable procedure.[82,83] These factors all reduce the efficacy of measurements made very close to vibrating surfaces. One positive factor in favour of close measurement is that the normal intensity component of extraneous fields is suppressed by reflection at the surface.[84]

It should be noted here that a considerable number of apparently successful determinations of source sound power and its surface distribution have been made using very close measurements on internal combustion engines. One probable reason for success is that engine blocks are very stiff, non-uniform structures which radiate rather efficiently at frequencies above about 200 Hz, in which case the active intensity component is strong and the reactive component is relatively weak. Also, the source radiates a high density of harmonic components, because the fundamental period of the mechanical generation process is long, and hence multi-frequency 'smearing' of circulatory flows of active intensity, mentioned earlier, eases the spatial sampling problem, particularly since results are usually presented in $1/3$ octave bands, each containing many harmonics. However, in cases of inefficient radiation, such as that from thin uniform plates vibrating far below their critical frequencies, near field effects can make close measurements virtually useless.[83]

In cases where adjacent regions of positive and negative normal intensity are present, a 'rule of thumb' for avoiding the recirculation regions is to withdraw the probe to a distance greater than the typical width of a surface region of uniform sign: this distance will, of course, vary with frequency.

(b) Geometric near and far fields

The far field directivity of a spatially extended source may be explained in terms of interference between the waves radiated from its various elemental regions. By definition, in the *far* field, the influence of differential spherical spreading on the resultant sum of the elemental fields is of minor importance compared with the delay (phase difference) effects of the individual path length differences. It is thus clear that the extent of the geometric near field is governed by the spatial extent of the (correlated) source region, since differential spreading is only significant if this region subtends a large angle at the observer position. In some cases, such as that of the ideal compact dipole, the geometric near field may not extend as far as the hydrodynamic near field, but this is not generally the case in practice.

It is clear that the relationship between pressure and particle velocity in geometric near fields of extended sources is an extremely complex function of position, even where the source itself is simple, such as the uniformly vibrating piston. The associated mean intensity field is correspondingly complex, as shown by Figs 8.1(a) and (b).

Non-uniform directivity is evidence of the presence of a finite-sized region of correlated source activity. Directional fields must be to some degree reactive because they are characterised by high spatial gradients of mean square pressure, at least in directions normal to a radial co-ordinate centred in the correlated region.[38] The simplest idealised case of a spatially extended source is that in which all the elemental (point) sources are uncorrelated, such as that used to represent a line of road traffic. The intensities of each source field simply add vectorially and the field is non-directional.

Although it is impossible to generalise about the pressure-intensity index in the geometric near field, it is clear that it will generally be positive, and sometimes quite large, especially if the probe is injudiciously oriented. As the measurement distance from the source increases, δ_{pI} for radially directed intensity measurements will generally decrease in the absence of extraneous sources or reflecting boundaries. With these present, and when all the significant sources emit broad band noise, it is likely that δ_{pI} will go through a minimum, in which case measurement bias errors will be minimised by

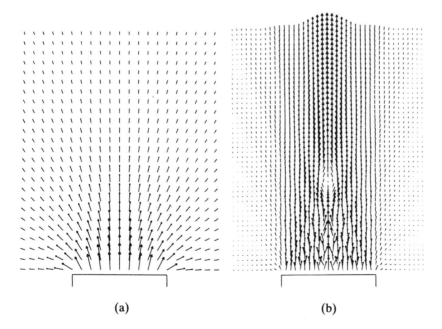

(a)   (b)

Fig. 8.1. Mean intensity field of a vibrating baffled piston of radius a; (a) $ka = 2$; (b) $ka = 25$.

selecting a measurement surface at this distance.[85] Where sources are tonal in character, such optimisation is not usually possible.

In the geometric far field, under free field conditions, the particle velocity and intensity vectors are purely radially directed, $I = \overline{p^2}/\rho_0 c$, $\delta_{pI} = 0$, and there is no need for intensity measurement. Where measurements are made in the combined far fields of a primary source and uncorrelated extraneous sources, δ_{pI} depends upon the relative directions and magnitudes of the component intensity vectors at the measuring point. For example, on some surface intermediate to the primary and extraneous source regions the normal intensity component will be zero at common source frequencies. This fact may be easily demonstrated with two loudspeakers fed with independent signals. Relatively small displacements of the measurement point along the common source axis produce large changes in the pressure-intensity index. It is clearly advisable to avoid such regions of a field, if possible.

The above considerations suggest that a rectangular, box-like measurement surface is likely to be inferior to a conformal surface, because L_{In} will tend to be substantially smaller than L_I in regions towards the edges of the box, thereby adversely affecting δ_{pI}, and measurement accuracy.

(c) Reflected wave fields

Here we shall consider waves which are radiated by sources to be reflected by large, uniform, planar surfaces, such as floors and walls. In practice, other more scattered and diffused reflections will be produced by irregular bodies in the field, but these generally cause less trouble than the former. Many sources operate on, or near, a rigid ground surface. For the purpose of acoustic analysis, the ground may be replaced by a coherent inverted image of the source. The real and image sources together form a combined source, and it is not sensible to attempt to separate them; they are *the* source. If a source does not normally operate near such a surface, it should not be tested near one.

Pure-tone intensity fields in enclosures generally exhibit circulatory patterns of active intensity, the spatial distributions of which are very sensitive to small alterations in source position, enclosure geometry, disposition of objects in the space (including people) and air temperature. Since all the source images are fully coherent, there is no far field, and the source power depends upon its position in the enclosure. The pressure-intensity index varies greatly with position, and strongly reactive regions of intensity exist. The special case of the one-dimensional reflective field has been analysed earlier in Chapters 3 and 4, in which it was shown that

$$I = (A^2/2\rho_0 c)(1 - R^2)$$

and

$$\overline{p^2} = (A^2/2)[1 + R^2 + 2R\cos(2kx + \theta)]$$

Hence, $(\overline{p^2}/\rho_0 c)/I = [1 + R^2 + 2R\cos(2kx + \theta)]/[1 - R^2]$, of which the maximum and minimum values are respectively $(1 + R)^2/(1 - R^2)$ and $(1 - R)^2/(1 - R^2)$: the range of δ_{pI} is seen to depend only on the magnitude of the reflection coefficient R. A qualitatively similar dependence on boundary reflection/absorption occurs in three-dimensional, reverberant, pure-tone fields; but, as mentioned earlier, intensity direction varies with position, thereby greatly increasing the range of measured δ_{pI} for any single probe orientation.

In the face of all these adverse factors, it is the author's contention that attempts accurately to determine by intensity measurement the sound power of pure-tone sources in reverberant enclosures are fraught with considerable uncertainty. Nor should it be forgotten that using a very-narrow-band signal analysis (e.g. FFT zoom) is tantamount to creating a very-narrow-band source, irrespective of the actual source bandwidth. In reverberant source environments it is therefore inadvisable to attempt to look too closely into spectral detail.

In cases where all sources present generate reasonably broad band noise, the measurement task is considerably eased. This is not because the various phenomena associated with narrow band fields are not operative, but because the frequency band integration (or averaging) process moderates their severity. In a large enclosure, which has a high density of acoustic mode natural frequencies, the concept of a diffuse field is tenable, and incoherence between direct and reflected fields may be assumed (see Section 4.8.5). The net intensity in an ideal diffuse field is zero (which is one reason why it can never be exactly realised) and therefore it influences δ_{pI} only through L_p. Since I decreases with distance from a source, δ_{pI} increases with distance from a source operating in a diffuse field (either self-generated or extraneous), and, in principle, an optimum intensity measurement surface which minimises δ_{pI} due to the combined effects of near and reverberant fields may, in principle, be found (Fig. 8.2: see also Section 9.3.4). In practice, such an optimum distance is not always evident, especially with narrow band or tonal sources in reverberant enclosures.[85]

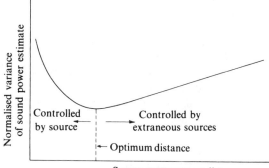

Fig. 8.2. Ideal optimum distance of a sound power measurement surface from the surface of a source.

8.4   CLASSIFICATION OF SOUND FIELDS: FIELD INDICATORS

Jacobsen[86] describes a sound field indicator as 'a normalized quantity that expresses some local or global property of the sound field'. The most widely used sound field indicator is the pressure-intensity index δ_{pI}, which is defined to be the decibel difference between the measured levels of sound pressure and sound intensity. It is expressed by eqn (4.55a) as

$$\delta_{pI} = 10 \lg[\overline{p^2}/I_r] - 10 \lg[\rho_0 c] \qquad (8.1)$$

It should strictly be denoted by δ_{pI_r} since I_r represents the magnitude of the component I_r *measured in the direction of a probe axis*, rather than that of the total mean intensity vector (which are generally not the same): however, typographical difficulties have caused the subscript r to be omitted in most publications. This quantity provides some general indication of the state of the sound field at the point of measurement; but in cases where it differs markedly from zero it gives no specific indication of the physical reason for the divergence. Among the possible reasons are the following:

 (i) the probe axis is nearly perpendicular to the direction of the mean intensity vector;
 (ii) the field is partially reactive; for example, in the hydrodynamic near field of a source or in a multiply reflected, coherent (interference) field;
(iii) the field is effectively 'diffuse'; for example, not very close to an individual source in a large factory building containing many independent sources;
(iv) the field is produced by two or more sources which generate oppositely directed intensity vectors of similar magnitude at the point concerned, which almost cancel each other;
 (v) the measurements are made in a frequency band in which the direction of the intensity vector varies in such a way that the magnitude of the band integral vector is small (i.e. the sign of the intensity alternates rapidly with frequency); for example, in the hydrodynamic near field of a vibrating panel which has many natural frequencies in the band,[83] or in a multi-mode reverberant field in an enclosure.

In order to distinguish between these different field states Jacobsen[86] has proposed various field indicators. It should be noted carefully that these may be based either upon pure-tone quantities or upon frequency-band integral (or average) values. The distinction is important because the sign of certain quantities can alternate rapidly with frequency, in which case their frequency-band integral (or average) values can be very small, even if they are not small at any one frequency.

8.4.1   Pressure–velocity coherence

A potentially valuable indicator of both the state of a sound field, and whether one or more independent sources is generating that field, is the coherence between the sound pressure and the particle velocity. This is given by

$$\gamma_{pu}^2(\omega) = \frac{|S_{pu_x}(\omega)|^2 + |S_{pu_y}(\omega)|^2 + |S_{pu_z}(\omega)|^2}{S_{pp}(\omega)[S_{u_x u_x}(\omega) + S_{u_y u_y}(\omega) + S_{u_z u_z}(\omega)]} \tag{8.2}$$

in which the symbol u_i represents the magnitude of one Cartesian component of the particle velocity vector, and the cross-spectral density S_{pu} is defined by eqn (5.10). Most intensity measurements employ single axis probes which sense only the component of particle velocity directed along the probe axis; consequently, a more useful quantity is the coherence between the sound pressure and a particle velocity component in a specific direction. In terms of the component of particle velocity u_r measured along the axis of a p-p probe, the expression reduces to

$$\gamma_{pu_r}^2(\omega) = \frac{|S_{pu_r}(\omega)|^2}{S_{pp}(\omega)S_{u_r u_r}(\omega)} = \frac{C_{pu_r}^2(\omega) + Q_{pu_r}^2(\omega)}{S_{pp}(\omega)S_{u_r u_r}(\omega)} \tag{8.3}$$

in which the cross-spectral density can be computed from eqn (5.21). As with all other quantities computed from the signals from a p-p probe placed in one orientation, the value of the quantity depends upon both the nature of the field and the specific orientation of the probe axis.

If a sound field is generated by a single (coherent) source, the *pu* coherence must be unity everywhere and at all frequencies: this is also the case where a number of physically separate pure-tone sources are generating identical frequencies, because they are fully correlated. In the hydrodynamic near field of any source, the pressure and particle velocity at any point are both principally controlled by the action of the source in the immediate vicinity of that point; therefore *pu* coherence tends to be close to unity, irrespective of the complexity of the source. In the geometric near field, on the other hand, significant contributions to the pressure and particle velocity are made by an extended region of the source system. It can be shown[87] that the *pu* coherence at a point in a field generated by two or more incoherent sources is less than unity if the acoustic impedances of the waves radiated from each independent source are different at that point. Consequently, unless the whole source region is fully coherent, the *pu* coherence falls below unity in the geometric near field; the actual value will depend upon the orientation of the probe axis. By definition, δ_{pI} is greater than zero in the geometric near field of any source, fully coherent or not. In the far field, if it exists, $p(t) = \rho_0 c u(t)$, irrespective of the nature of the source, and the *pu* coherence is unity. This suggests that the value of the coherence between the pressure and the component of particle velocity directed radially out of the source 'centre' may be used as an indicator of the extent of the geometric near field of a complex radiator, provided that no significant reflections (image sources) exist. If a source normally operates in close proximity to a reflective surface, such as a floor, the source and its coherent image cannot be considered as independent; they act in concert as a single source.

In the ideal 'diffuse' field, described in Section 4.8.5, the *pu* coherence is zero. A close approximation to such an ideal field would be generated by a large number of independent machines in a reverberant factory space, located at points well removed from the observation region. The reader is here reminded to be eternally vigilant when measuring coherence with a signal analyser; a common cause of apparently poor coherence is bias error due to inadequate frequency resolution. This source of error is particularly virulent in sound fields generated within highly reflective enclosures, because the period during which the waves emitted by a source continue to bounce around the enclosed volume may be far longer than the time sampling window of the analyser associated with the selected frequency range, and the bandwidth of the acoustic modes tends to be smaller than a typical analysis resolution bandwidth; in other words, the 'acoustic memory' of the room is longer than that of the analyser. The crucial test is to reduce the frequency range and observe the computed coherence; ideally, the resolution bandwidth should be much less than $2 \cdot 2/T$, where T is the enclosure reverberation time. An example of the application of *pu* coherence measurements to the characterisation of the sound field around a large machine is presented in Section 9.5.

Jacobsen[79] has also proposed a finite frequency band coherence which is proportional to the square of the modulus of the integral of the *pu* cross-spectrum over the defined band. Note that this is different from the frequency band *pu* correlation coefficient which is proportional to the integral over a band of the cross-spectrum (see eqn (5.11)). If the *pu* phase varies widely over the range $0-2\pi$ within the band, the frequency band correlation coefficient will be far less than unity, and will tend to zero for a random phase distribution, even if the coherence within the band is unity. The computed value may, of course, be biased, as indicated above. Figure 8.3 shows an example of low coherence associated with the action of two independent sources; note that the coherence is effectively independent of the frequency resolution of analysis. By contrast, Fig. 8.4 shows the profound effect of resolution on the coherence measured in a field generated by a single monopole; it is, in reality, unity at all frequencies. It is seen therefore that the sensitivity of computed coherence to spectral resolution provides an additonal indication of the nature of a sound field.

8.4.2   Active and reactive field indicators

In Section 4.4 it was shown that in a plane progressive wave the mean intensity is related to the mean energy density by

$$I = ce \tag{8.4}$$

where c is the group speed which, because sound waves are non-dispersive, is equal to the phase speed: the intensity of such a wave is purely active. It has been suggested[86,88] that the degree of conformity with this relationship could be taken as a measure of the 'activeness' of a field, and an associated indicator

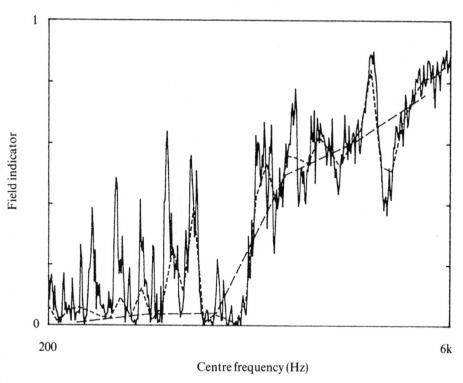

Field indicator

1

0

200

6k

Centre frequency (Hz)

Fig. 8.3. Frequency-band coherence $\gamma^2_{pu}(\omega, \Delta\omega)$ measured in a sound field produced by two uncorrelated sources. Spectral resolution: —— 15 Hz; ---- 150 Hz; — — 1 kHz.[86]

defined by

$$\psi_r = 10 \lg(ce_r/|I_r|) \tag{8.5}$$

has been proposed. Note that the energy density and intensity symbols carry a subscript r, denoting evaluation on the basis of the component of the particle velocity in direction $\mathbf{r}: e_r$ is not the total energy density. The spectral value of e_r is equal to the sum of the spectral components of its potential and kinetic components which may be computed from the signals from a p-p probe having sensor separation d as follows:

$$G_{e_{pot}} = G_{pp}/2\rho_0 c^2 = [G_{p_1 p_1} + G_{p_2 p_2} + 2\,\mathrm{Re}\{G_{p_1 p_2}\}]/8\rho_0 c^2 \tag{8.6a}$$

and

$$G_{e_{kin,r}} = \rho_0 G_{u_r u_r}/2 = [G_{p_1 p_1} + G_{p_2 p_2} - 2\,\mathrm{Re}\{G_{p_1 p_2}\}]/2\rho_0 \omega^2 d^2 \tag{8.6b}$$

This indicator compounds the effect of probe orientation with the state of the field. It is similar to δ_{pI} which may be written as

$$10 \lg[2ce_{pot}/I_r]$$

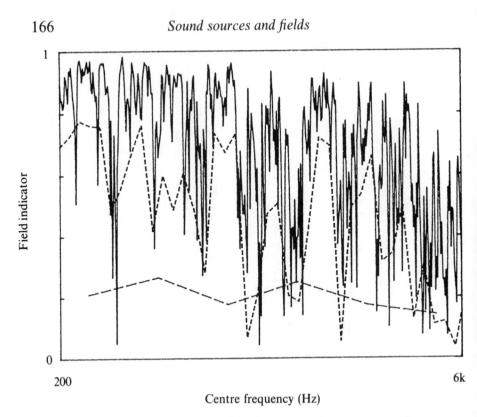

Fig. 8.4. Frequency-band coherence $\gamma_{pu}^2(\omega, \Delta\omega)$ measured near a monopole source in the presence of strong reflections. Spectral resolution: —— 15 Hz; ---- 150 Hz; ——— 1 kHz.[86]

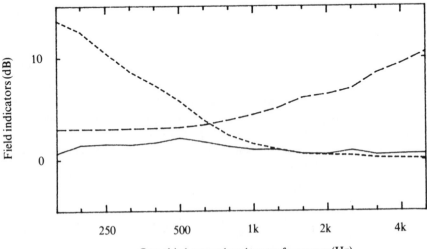

Fig. 8.5. Measurements in the near field of a monopole source: —— pressure-intensity index $\delta_{p\|I\|}$; ---- energy density-intensity ratio ψ_r; ——— near field indicator Γ_r.[86]

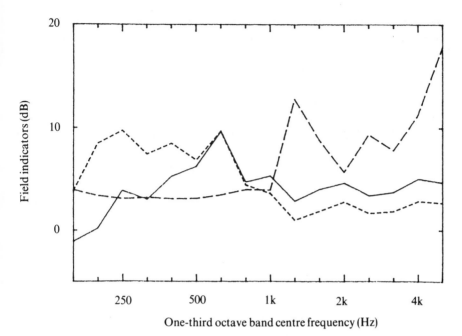

Fig. 8.6. Measurement in the near field of a vibrating plate: —— pressure-intensity index $\delta_{p|I|}$; ---- energy density-intensity ratio ψ_r; ––– near field indicator Γ_r.[86]

This proposed indicator has a basic weakness as a field indicator, which is related to that of δ_{pI}. One would assume that if the 'activeness' indicator were very different from zero, the field would necessarily be strongly reactive; this is not, however, necessarily the case. If a measurement is made at a position situated between independent sources of similar sound powers, the oppositely directed intensity components vie for dominance (and will totally cancel each other at some point), but the component energy densities will add: ψ_r will be large, although the field is totally non-reactive. An ideal 'diffuse' field created in free space by a spherical array of uncorrelated sources is totally non-reactive, but ψ_r will be, in principle, infinite. Hence ψ_r indicates how close a field is to a freely propagating far wave field, but alone is not a sufficient gauge of the lack of 'reactivity' of a sound field.

Of somewhat greater potential utility is a proposed companion indicator Γ_r, which compares the reactive intensity in a field to that in a purely reactive field. It may be shown that the sum of the squares of the active and reactive intensity components in any direction in a *pure-tone* sound field equals $4c^2 e_{pot} e_{kin,r}$, where e_{pot} is the potential energy density and $e_{kin,r}$ is the kinetic energy density associated with direction **r**. Indicator Γ_r is defined by

$$\Gamma_r = 10 \log[2c \sqrt{(e_{pot} e_{kin,r})}/|J_r|] \tag{8.7}$$

where $|J_r|$ is the magnitude of the component of the total reactive intensity associated with particle velocity component u_r. The energy density components can be determined from the output from a p-p probe, as indicated by eqns (8.6). Measured values of ψ_r and Γ_r near a monopole and a vibrating plate are compared with δ_{pI} in Figs 8.5 and 8.6. Note again that this indicator is not useful in purely active fields at positions where the net active intensity is low, or zero, because of the presence of opposing, uncorrelated wave field components having similar active intensities.

It has been suggested[89] that a related global indicator

$$\Delta_{JI} = 10\lg\left|\int_S \mathbf{J}.\mathrm{d}\mathbf{S}\middle/\int_S \mathbf{I}.\mathrm{d}\mathbf{S}\right| \tag{8.8}$$

evaluated on a measurement surface enclosing a source region, could be used to establish the extent of the near field. The vector reactive intensity $\mathbf{J}$ is given by[90]

$$\mathbf{J} = c\nabla e_{\mathrm{pot}}/k \tag{8.9}$$

CHAPTER NINE

Determination of sound power using sound intensity measurement

9.1 WHY SOUND POWER?

Radiated sound power is widely considered to be an intrinsic property of a source which may be used as a quantitative 'label' of acoustic output. Although it is true that source sound power is generally far less influenced by the nature of the immediate surroundings than the associated sound pressure, it is not totally independent of these surroundings, as illustrated in Sections 9.2 and 9.5. The principal practical importance of sound power is that it is the fundamental quantity upon which a prediction of the potential effect of a source on any operational environment must be based. In addition, it constitutes the basis for comparison between the acoustic performance of products from different suppliers; it is one of the two major acoustic quantities which are controlled by legislation and regulations; and its dependence upon the physical and operational parameters of a noise-generating system provides a basis for source identification and characterisation.

Although the quantity of principal concern to those who attempt to control noise is sound pressure, since it is the agent of hearing damage and the other adverse effects of noise, it is virtually impossible to estimate the precise sound pressure generated by a source at any specific point in its sound field. In the open air, micrometeorological phenomena such as local air temperature distributions and the mean and fluctuating airflows of wind, together with the topographical and acoustic properties of the ground, influence the propagation of sound energy; even if free field source directivity is known, predictions of the resulting sound pressures produced at distances of more than a few tens of metres carry significant uncertainty. The sound pressure distribution generated by a source operating in an enclosure is affected by reflection, scattering, diffraction and absorption by the enclosure boundaries and distributed objects. Hence, because source sound power is far less dependent upon the operating environment than the radiated pressure, the development of reliable methods for its measurement is of great practical importance. The advent of sound intensity measurement has greatly advanced this area of acoustic technology.

The principal applications of sound intensity measurement may be broadly classified as follows:

  (i) determination of the sound power of sources;
  (ii) measurement of the transmission of sound energy through partitions;
 (iii) measurement of the sound absorption properties of materials and structures;
 (iv) identification and rank ordering of source regions;
  (v) measurement of sound energy flow in waveguides.

A factor which is common to all these applications of sound intensity measurement is the determination of the sound power passing through some selected surface within a fluid. Even in the process of source location, it is essentially the sound power of source regions, rather than the sound intensity (sound power flux *density*) which is of significance. For example, a small leak in a partition may produce local intensities far in excess of the average over the remainder of the partition surface, but the area of the leak may be so small as to render negligible its contribution to the total transmitted sound power. An implication of considerable practical importance is that, in cases where the normal sound intensity component varies widely with position, errors incurred in estimating local intensity must be considered in relation to the associated *area-weighted* contribution to the total sound power passing through the measurement surface. It is therefore just as important to minimise errors in the estimation of low intensities distributed over large surfaces as it is in cases where high intensities exist over relatively small areas.

9.2   PRINCIPLE OF SOURCE SOUND POWER DETERMINATION

The energy conservation equation in the form of eqn (4.11a) states that, in a steady sound field, the divergence of the mean intensity vector equals the local volumetric density of the mean rate of generation of sound power. (Henceforward the word 'steady' will be used to mean 'stationary in time'.) Integration of this equation over a finite volume may be accomplished by the use of Gauss's Integral Theorem, by which the volume integral of the divergence of any field vector may be expressed in terms of the integral over the enclosing surface of the component of the vector normal to the surface:

$$\int_V \nabla . \mathbf{A} \, dV = \int_S \mathbf{A} . \, d\mathbf{S} = \int_S \mathbf{A} . d\mathbf{S}\mathbf{n} = \int_S A_n \, dS \qquad (9.1)$$

where $\mathbf{n}$ is the unit normal vector of the surface S which encloses the volume V, and A_n is the magnitude of the component of vector $\mathbf{A}$ normal to the surface.

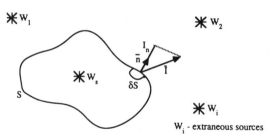

Fig. 9.1. The surface integral of I_n equals W_s; all other W_i excluded.

Hence, from eqn (4.11a)

$$\int_V \nabla.\overline{\mathbf{I}(t)}\,dV = \int_V \nabla.\mathbf{I}\,dV = \int_S \mathbf{I}.\mathbf{n}\,dS = \int_S I\cos\alpha\,dS = \int_S I_n\,dS$$

$$= \int_V \overline{W'(t)}\,dV = W_S \qquad\qquad (9.2)$$

where W_s is the net mean sound power generated by source mechanisms in their operation on the fluid contained within the enveloping surface S (Fig. 9.1). The qualification 'net' is necessary because there might exist within V negative source mechanisms of sound power absorption or dissipation. Normal components of sound intensity generated by *steady* source mechanisms operating on fluid *external* to S do not contribute to the surface *integral*, and hence to W_s, but they do add vectorially to the local intensity generated by the enclosed sources, thereby altering the surface distribution of the normal sound intensity component. The integral only applies if all sources, both internal and external to S, are steady. Unsteady cases, such as transient sound sources, can be similarly treated, but time must enter as an additional variable of integration (see Section 10.6).

It is sometimes believed that normal intensity components produced on surface S by reflections from surfaces outside S of the waves radiated by the enclosed sources, should be excluded from the integral. Under steady conditions, this is not so. If the enclosed sources do not absorb (dissipate) any energy transported into V by such reflected waves, it is all subsequently transmitted/scattered out through S, just as that due to active external sources, thereby contributing nothing to the net power flow through S. If, however, the enclosed sources do dissipate part of this reflected, or externally generated energy, the surface integral is diminished; in this case, the apparent power of the enclosed sources depends upon the effectiveness of the sound absorption mechanisms, the reflectivity of the source environment, and the sound power incident upon the enclosed absorbing regions from external sources. The integral theorem remains valid, but the effect is to alter the estimated source sound power from its free field value. In practical determinations of source sound power the presence of such absorption can constitute a serious problem.

An extremely important corollary of eqn (9.2) is that, for the purpose of source sound power determination, the choice of the enveloping surface *defines* 'the source', irrespective of the apparent physical and geometric characteristics and boundaries of the system concerned.

The source mechanisms operating in a volume V enclosed by a surface S may owe their existence to, or be influenced by, events occurring in systems external to S—for example, structural surface vibration generated by a remote mechanical source. The fluid within S 'knows' nothing about the transport through S into V of vibrational energy within the structural waveguide, except through any acoustic radiation effect it might have at the interface between the two, in which case the structure acts as a source of sound power. The vibrational energy entering the volume in this way may be partly dissipated by structural damping mechanisms, and may partly leave again via the structural path; therefore, it is not included in, but may affect, the surface integral of sound intensity which accounts only for energy lost from the structure by acoustic radiation within S. (An equation analogous to (4.11b) can be written for vibrational wave energy within a structure; see, for example, Ref. 91.) On the other hand, sound energy generated in the fluid within S may enter a structure within S and leave the volume via a structural path in the form of vibrational energy; the surface integral correctly takes no account of this power flow, which is inherent in the source system. However, mechanical structures within S can also act as absorbers of sound power generated by sources external to S, thereby reducing the apparent source power as explained above.

Although neither sources nor reflective surfaces external to S directly contribute sound power to W_s, they can, in principle, influence W_s. This seemingly paradoxical fact does not in any way invalidate eqns (4.11a) or (9.2). We have already seen how the presence of sound absorption within S can have this effect. There is another mechanism of influence which is more fundamentally related to the process of generation of sound power by sources. Sound waves falling on a source, which are coherent with (phase related to) the source action, affect the sound power generated by that action. A simple, but striking, example of this phenomenon—the monopole near a reflecting plane—has already been cited. The mean sound power generated by such a Category 1 (volumetric) source is equal to the time-average product of the pressure at the source point with the source volume velocity. Any agent which alters this pressure alters the sound power. If the agent is another coherent point source, the resulting change of power will be exactly reflected in the change in the value of the integral of normal intensity over any closed surface which encloses the 'primary' source, but excludes the interfering source. The power radiated by this latter source is totally excluded by this process of integration. On the enveloping surface S, the total pressure is, by the principle of superposition, the sum of the pressures generated by the two sources *in isolation*; similarly the two

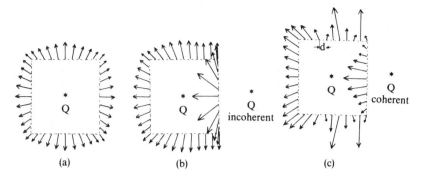

Fig. 9.2. Mean intensity vectors on a measurement surface: (a) an isolated harmonic point monopole; (b) two harmonic monopoles of different frequencies (incoherent); (c) two harmonic monopoles of equal frequency ($kd = 1.83$).

particle velocity vectors add vectorially. Consequently, the surface intensity has contributions from four components, two being those of the individual isolated sources, and two being the result of interference between the two source fields. Not only is the integral changed in value by the presence of the interfering, source, but the surface distribution of normal intensity is totally altered. This phenomenon is illustrated in Fig. 9.2(a) which shows the intensity distribution from an isolated monopole, and Fig. 9.2(b) in which the sources are of equal strength but different frequency (therefore incoherent). In this case the total intensity vector at any point is the vector sum of the intensity vectors of the individual isolated sources. The distribution of normal intensity is altered by the presence of the other source, but the surface integral is unchanged. In the case illustrated in Fig. 9.2(c), the second source is fully coherent with the first; as a result, both the intensity distribution and the surface integral of normal intensity are altered by the presence of the second source; the power radiated by the first source is therefore also altered. A practical example of this phenomenon is shown in Fig. 9.3.[38]

Reflections by large plane surfaces, such as floors and walls, of waves radiated by sources, may be replaced in theoretical analysis by image sources, just like optical images in flat mirrors (e.g. Fig. 3.10). These image sources are fully coherent with the primary sources which they reflect; consequently such reflecting surfaces affect the power output of sources in their vicinity. The effect diminishes with distance of the reflector from the source because of the inverse square law.

The influence of coherent interfering sources and reflections is most marked where primary sources have strong tonal, or narrow-band, components. The frequency-average effects on broad band sources tend to be weaker because *correlation* between radiated and reflected pressures tends to decrease with increase of bandwidth even if they are fully coherent. However, if the sound intensity and power analyses are made in narrow bands, the effects will be

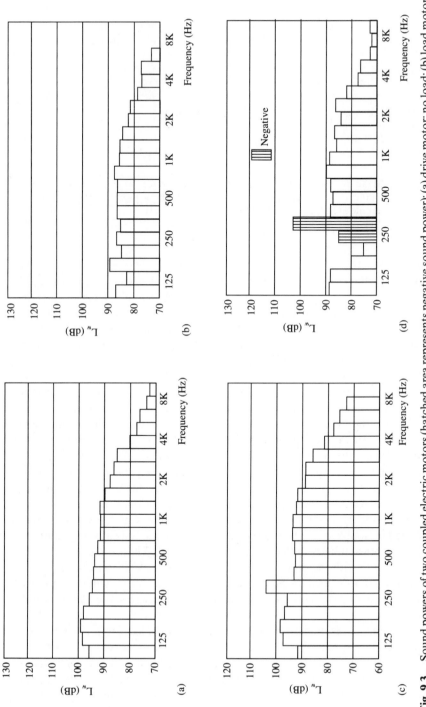

Fig. 9.3. Sound powers of two coupled electric motors (hatched area represents negative sound power): (a) drive motor: no load; (b) load motor: passive; (c) drive motor: load motor active; (d) load motor: active.[38]

evident, even with broad band sources. An example of the influence of reflecting surfaces on source sound power is presented in Section 9.5.

At the risk of insulting the intelligence of my readers I feel that I should re-emphasise the fact that no intensity measurement system is a 'magic' device which can somehow discriminate between the component of sound intensity that one wishes to measure (say the outgoing waves radiated by a machine in a factory), and the contributions to the intensity field from any other source present, or from reflections from surrounding obstacles of the outgoing waves from the primary source. A probe cannot, and indeed should not, discriminate; it measures *net* intensity. It is only in the implementation of the surface integral expressed by eqn (9.2) that the total sound power of a chosen source is selected out, and all other contributions are nullified. For this reason, the directivity of a source cannot be determined by sound intensity measurement in any environment other than free field, in the absence of other sources; and then only if measurements are made of the normal component of intensity on a spherical surface in the far field of the source.

Another point of detail which seems not to be universally understood is that a sound power measurement surface must *completely* enclose the source under investigation, even if that surface lies close to the source surface; the edge boundaries must be closed and the associated intensity distribution evaluated. For example, measurements of the normal intensity component on subareas of a plane surface lying parallel to a window pane are not sufficient for the purpose of rank ordering the transmission through the various subareas; sound energy can, and often does, travel along paths nearly parallel to the radiating surface, and can therefore emerge at a quite different position from its source. Of course, there are many situations in which a source is located on, or in close proximity to, a solid surface which is known to be acoustically inactive, in which case the measurement surface may incorporate a portion of that surface, on which measurements need not be made.

9.3 SOURCES OF ERROR IN SOUND POWER DETERMINATION

As shown in the previous section, the sound power radiated by any continuous, *steady*, sound-generating system (or part thereof), operating in the presence of other *steady* sources, may, in principle, be obtained by enclosing that system (or part thereof) by an imaginary enveloping surface and evaluating the surface integral of the component of sound intensity normal to the surface. The principal sources of uncertainty and error in the practical implementation of this principle, apart from temporal variations in the acoustic environment and/or in the outputs of significant sources, are: (i) intensity field spatial sampling error; (ii) instrument bias error associated with phase mismatch; and (iii) random error associated with temporal averaging and spectral estimation. These are of central concern in the development of

Table 9.1 Errors in sound power determination with p-p intensity probes[92]

Source of error	Type	Predictable from measurable indicator	Correctable	Error inversely proportional to Δr	Frequency range
Finite difference	Bias	No	No	No	High frequency
Phase mismatch	Bias	Yes	Yes	Yes	
Low vent attenuation	Bias	Yes	Yes	Yes	Low frequency
Spatial sampling	Random[a]	No	No	No	
Finite averaging time	Random[a]	Yes	No	No	
Electrical noise	Random	Yes	No	Yes	Low frequency
Non-stationarity	Random[a]	No	No	No	
Airflow fundamental effect	Bias	No	No	No	
Turbulence	Bias + random	No	No	Largely	Low frequency
Effect of windscreen losses	Bias[a]	Partly	Partly	No	Low frequency

[a] Not related to the measurement principle.

standardised methods of sound power determination of which a vital component is the specification of confidence ranges for the results obtained by application of the standards.

Although the principal purpose of normal intensity sampling for sound power determination is to provide an estimate of the surface integral, or area-weighted mean value, the spatial distribution of normal intensity over the measurement surface is also of concern because it influences the accuracy of, and therefore confidence in, the sound power estimate. Naturally, only the sample distribution, and not the true distribution, can be determined.

Table 9.1[92] summarises the sources of error in sound power determination.

The following sections treat the origins, forms and significance of the various errors to which sound power determination based upon intensity measurement is subject.

9.3.1 Normal intensity field sampling error

It is, of course, physically impossible to implement exactly the integral expressed by eqn (9.2) because it covers every point on the measurement surface; consequently, the integral must be approximated by sampling the normal intensity field at selected points, or along selected paths, to estimate its area-weighted mean value. The application of any sampling procedure, whether to a set of discrete variables (such as the heights of forty-year-old people in Brazil), or to a continuously distributed variable (such as the height above sea level of the ground surface in the Netherlands), is very likely to produce estimates of statistical descriptors, such as mean value and variance, which differ from the true value: this constitutes sampling error. The error of any one estimate derived from a finite sample cannot, of course, be known (since the true value is not known). The magnitude of this error is uncertain; if it is likely to be small, the estimate is valid and useful; if it is likely to be large, the estimate is misleading and not useful.

If the sampling procedure were such that individual samples were statistically independent, and if it were to be repeated many times in such a manner that the resulting estimates were statistically independent (e.g. not based on samples which were strongly correlated with those of another sample set), the set of estimates so produced would possess rather special statistical properties. According to the Central Limit Theorem the set of estimates of mean values would have a normal (Gaussian) distribution and a variance equal to the *true* variance of the sampled variable divided by the size of the individual sample set, irrespective of the form of distribution of the variable. In this case, the probability of any individual estimate of the mean value of a variable lying within a specified range of the true value can be quantified in terms of confidence limits. There is one major snag—the confidence limits are determined by the true variance of the sampled variable, which is *a priori* unknown! This 'Catch 22' situation may be resolved either by the assumption of a plausible model of the distribution of the variable, or that the variance of the measured sample set approximates to the true variance—a kind of 'boot-strapping'. It is intuitively obvious that this approximation improves with the size of the sample set and with the 'smoothness' of the distribution of the variable.

Sound intensity distributions are inherently more difficult to characterise than distributions of mean square pressure because intensity bears a sign in addition to magnitude and strongly bipolar distributions are common, especially in the presence of strong extraneous sources.[93] This situation can lead to large uncertainties in the evaluation of source sound power because the net power is derived from a difference between the estimates of two quantities of similar magnitude (the outward- and inward-flowing powers), each bearing a degree of uncertainty which can be of the same order of magnitude as the true net power.

Two sampling techniques are currently practised in sound power determination using sound intensity measurement. In the 'discrete point' method, the intensity probe is held stationary at each of a number of fixed points on a measurement surface for a period of time which is sufficiently long to make the temporal averaging error acceptably small. In the 'scanning' method, the intensity probe is moved continuously over a prescribed path on the measurement surface during which time the intensity measurement instrument performs a time-averaging process: time-averaging is thus substituted for space-averaging, the relationship being determined by the speed of probe motion, or 'scanning speed'.

The sampling error associated with discrete point sampling is related to the spatial distribution of normal intensity over the measurement surface and to the form and size of the sampling array employed. If it may be assumed that individual intensity samples are *statistically independent*, estimates of the average intensity (or sound power) obtained by repeated random reorientation of the array will, according to the central limit theorem, be normally distributed, and the associated variance will be equal to the variance of the actual intensity distribution divided by the size of the sample. Hence the confidence that a single estimate of sound power lies within any specified range about the true value may, in principle, be estimated. International Standard ISO 9614-1 implicitly employs this assumption of statistically independent samples to specify the number of intensity measurement points necessary to ensure that the error in the estimate of source sound power associated with the spatial sampling error is restricted to within an acceptable range at a specified level of confidence (see Section 9.4). It is also implicitly assumed that the number of samples is sufficient to ensure that their variance provides a good (or, at least, conservative) approximation to the true spatial variance of the normal intensity.

The question of the degree of statistical independence of discrete point samples of normal intensity has not at present been resolved. Sources and their intensity field distributions are so diverse in character that no generally valid conditions can be proposed. The problem is compounded by the fact that the total intensity at any point in a sound field generated by multiple sources is determined by the co-operation of the sums of the sound pressures and particle velocities produced by each source at the field point. Hence, if a source operates in the presence of others, the spatial distribution of normal intensity over any measurement surface enclosing the source is not solely determined by the field generated by that source alone. It is clear on physical grounds that the spatial correlation scale of the source strength density must influence the spatial correlation of the free field intensity distribution. The direct radiation field of a source cannot be diffuse, or even hemi-diffuse, because the source subtends a finite solid angle of less than 2π steradians at any observation point: hence it is not appropriate to base any estimate of correlation scales on room acoustic models. It seems likely that, for an isolated source, sample independence is likely to decrease with distance from the source; this surmise has not been subjected to empirical test. In cases where samples are not actually statistically independent, the assumption of independence will demand more samples than necessary to achieve a given confidence limit; however, it is not possible to identify these cases on the basis of any known criteria, except that as sample point separations decrease the samples become increasingly correlated.

In the 'scanning' method, the probe moves continuously along a path on the measurement surface. Because the normal intensity varies with position, the uncertainty associated with the estimate of the average normal intensity over any small portion of the scan path can be high, especially for narrow band fields; the probe simply does not remain in the close vicinity of any one position long enough for the temporal averaging procedure to generate a reliable estimate. Spatial and temporal sampling errors are compounded. This problem is particularly severe in fields where the relative phase of the two pressure signals varies rapidly in spaces, for example in the near field of a thin vibrating panel.[83] If a source is highly directional, so that the radiated power is concentrated in relatively small regions ('hot spots') of the measurement surface, any error due to an excessively fast traverse of these regions may produce a serious bias error in the sound power estimate. Of course, this problem also plagues the discrete point sampling method in which the selected sampling points may 'miss' the hot spots. The signals received from two moving microphones are also non-stationary in time, so that the results of conventional Fourier analysis are not strictly valid; however, this does not seem to have any serious practical implications except at scan speeds greatly in excess of $1 \, \text{m s}^{-1}$ in very complex intensity fields such as those close to thin panels and in highly reverberant enclosures. The random error of sound power determination associated with scanned measurements is discussed further in Section 9.3.3, together with some examples of comparisons between the results of point-sampled and scanned measurements.

In a paper which presents a combined theoretical and experimental investigation of sampling error, Jacobsen[94] contends that point samples of normal intensity on a measurement surface cannot be considered to be statistically independent, and that therefore the variance of a power estimate will not necessarily vary inversely with the size of the sample (or even, necessarily, decrease with sample size in the case of highly correlated samples). He demonstrates that the sampling error in estimates of space-average intensity produced by uniformly spaced sample points is related to the degree of aliasing of the spatial frequency (wavenumber) spectrum of the intensity distribution produced by violation of the Shannon sampling theorem: error may be introduced if the array is too coarse to resolve small-scale spatial variations of relatively large amplitude. Unfortunately, most practitioners of sound intensity measurement for the purpose of sound power determination do not have the time to evaluate the wavenumber spectrum of the true intensity distribution which can, of course, only be estimated by substantial oversampling to suppress aliasing. Jacobsen also demonstrates, on the basis of intensity field sampling close to a loudspeaker and to a thin vibrating plate, that spatially random sampling produces variances in sound power estimates which vary with sample size in a manner expected of statistically independent samples. However, he concedes that non-randomly spaced sampling is far more practical, even though the sampling error is unpredictable.

Experimental results[94] also suggest that the assumption of statistically independent samples for estimating the number of samples necessary to restrict the associated error in sound power estimate to within a specified range may lead to a significant overestimate of the required number of measurement points, by a factor which can exceed ten under difficult measurement conditions. It would be dangerous to generalise this result because many practical sound power determinations are made under very adverse acoustic conditions: in addition, there exists a large body of empirical evidence to the contrary (e.g. Ref. 93). The experimental investigations also demonstrate that an insensitivity of sound power estimate to a doubling of sample size on a given surface is not an unequivocal indication of low sampling error, a conclusion which casts some doubt on the effectiveness of the sample size doubling test for convergence employed in ANSI Standard S.12.12, the North American counterpart of ISO 9614. It is not possible to estimate precisely the spatial sampling error associated with scanning because it is compounded with spectral estimation errors associated with time-averaging (see Section 7.3), and the nature of the sound field, which cannot be estimated from the end result of a scanned measurement.

As mentioned previously, where there is a fine balance between sound power passing outwards through one or more regions of a measurement surface (positive partial power) and that passing inwards (negative partial power) due either to the presence of external sources, or to near field intensity circulation, the sum of the likely ranges of errors of the estimates of these powers must be considerably less than the value of the difference if the difference is to have a statistical significance. For example, if the 95% confidence limits of both power estimates are $\pm 2\,dB$, then a difference of $3\,dB$ between the estimated positive and negative partial powers cannot be considered to be reliable at a confidence level of 95%. Note that the net power would be 3 dB below the positive partial power in this case. If the powers were 2 dB different, the net power would be approximately 4·5 dB below the positive partial power.

9.3.2 Instrument bias error

Phase mismatch between probe transducers and associated signal processor channels is the predominant source of systematic error in sound intensity measurement. The associated normalised error at any one frequency at any one position is equal to the ratio of the phase mismatch ϕ_s to the actual p-p phase difference ϕ_f in the sound field (see eqn (6.3) and Fig. 6.11). An estimate of source sound power based upon biased estimates of normal intensity on the measurement surface is likely to be biased. From eqn (7.1) the unbiased normal intensity estimate at any point on the measurement surface is given, within the accuracy of the finite difference approximation, by

$$I = (\overline{p^2}/\rho_0 c)\phi_f/kd \tag{9.3}$$

and the biased estimate is given by eqn (7.2) as

$$I_e = (\overline{p^2}/\rho_0 c)(\phi_f + \phi_s)/kd \tag{9.4}$$

in which it is assumed that the estimate of $\overline{p^2}$ is unbiased. The corresponding unbiased and biased estimates of sound power derived from scans of normal intensity and mean square pressure over a measurement surface are given by

$$W = (1/kd) \int_S (\overline{p^2}/\rho_0 c)\phi_f \, dS \tag{9.5}$$

and

$$W_e = (1/kd) \int_S (\overline{p^2}/\rho_0 c)(\phi_f + \phi_s) \, dS \tag{9.6}$$

Hence

$$W_e - W = (1/kd)\phi_s \int_S (\overline{p^2}/\rho_0 c) \, dS \tag{9.7}$$

When the microphones of an intensity probe are exposed to identical pressures in a calibrator, the residual intensity produced by phase mismatch is given by

$$I_{res} = (\phi_s/kd)(\overline{p^2}/\rho_0 c)_{res} \tag{9.8}$$

and hence

$$\phi_s = I_{res}(kd)/(\overline{p^2}/\rho_0 c)_{res} \tag{9.9}$$

Substituting for ϕ_s in eqn (9.7) gives

$$W_e - W = [I_{res}/(\overline{p^2}/\rho_0 c)_{res}] \int_S (\overline{p^2}/\rho_0 c) \, dS \tag{9.10}$$

and

$$(W_e - W)/W_e = [I_{res}/(\overline{p^2}/\rho_0 c)_{res}]\left[\int_S (\overline{p^2}/\rho_0 c)\,dS \middle/ \int_S I_e\,dS\right] \quad (9.11)$$

This is the normalised error based upon the biased estimate W_e.

We may define a measurement-surface-average (or 'global') pressure-intensity index as

$$\langle \delta_{pI} \rangle = 10\lg\left[(1/S)\int_S (\overline{p^2}/\rho_0 c)\,dS \middle/ (1/S)\int_S I_e\,dS\right] \quad (9.12)$$

in which S denotes the total area of the measurement surface. (This quantity is known as the Field Index F_{pI} in International Standard ISO 9614-2.) Surface-average, rather than surface-integral, values appear in eqn (9.12) because they correspond to the quantities produced by the practical scanning procedure. Hence the biased sound power estimate may be expressed as

$$W_e = \int_S I_e\,dS = \left[\int_S (\overline{p^2}/\rho_0 c)\,dS\right][10^{-\langle\delta_{pI}\rangle/10}] \quad (9.13)$$

The ratio of residual intensity to mean square pressure in a calibrator is usually expressed as the pressure-residual intensity index given by

$$\delta_{pI0} = 10\log[(\overline{p^2}/\rho_0 c)_{res}/I_{res}] \quad (9.14)$$

from which

$$W_e - W = [(I_{res})/(\overline{p^2}/\rho_0 c)_{res}]\int_S (\overline{p^2}/\rho_0 c)\,dS = 10^{-\delta_{pI0}}\int_S (\overline{p^2}/\rho_0 c)\,dS \quad (9.15)$$

Hence, from eqns (9.11), (9.13) and (9.15) the normalised bias error is given by

$$e_\phi(W) = [10^{(\delta_{pI0} - \langle\delta_{pI}\rangle)/10} - 1]^{-1} \quad (9.16)$$

and its logarithmic form is

$$10\lg[1 + 1/e_\phi(W)] = \delta_{pI0} - \langle\delta_{pI}\rangle \quad (9.17)$$

Comparison of these eqns with eqns (7.11b) and (6.3) indicates that the surface-average pressure-intensity index plays the same role in determining the bias error of sound power estimation as the point pressure-intensity index δ_{pI} does in determining the bias error in point intensity estimation. Local values of δ_{pI} may be such as to exceed the dynamic capability L_d of the measuring instrument, as determined by the desired maximum bias error of the determination (see Section 6.9.1); nevertheless, the bias error of the sound power determination will be acceptable if $\langle\delta_{pI}\rangle$ does not exceed L_d. Provided that the phase mismatch ϕ_s varies smoothly and slowly with frequency, eqn (9.17) may also be applied to measurements made in 1/3 octave frequency bands.[94]

The equivalent expression for the normalised bias error associated with discrete point measurements is obtained by replacing the surface integrals in eqns (9.12) and (9.13), by the sums of the measured point intensities and mean square pressures weighted by the measurement surface areas associated with each point.

Equation (9.15) indicates that the (unnormalised) bias error of a sound power estimate made with a given instrument (i.e. given δ_{pI0}) is proportional to the integral of the mean square pressure over the measurement surface. Since the true sound power is independent of the measurement surface under all conditions, the normalised bias error must also be proportional to this integral. In free field, in the absence of extraneous sources, this integral will generally decrease with distance from an extended source in the near field; it may increase in the geometric near field (see Section 8.3) and then stabilise once the geometric far field is reached. In a reverberant environment, the integral will tend to increase with distances from the source beyond the 'reverberation radius'. The presence of extraneous sources will increase the integral rapidly beyond distances where the surface-average mean square pressure of the source equals that of the extraneous source. There is likely, therefore, to exist in practice some measurement surface which minimises this integral; its approximate location can be established by means of rapid scans of mean square pressure over candidate surfaces (see Fig. 8.2). It must be remembered that a scan produces a line average which is assumed to approximate to the surface average; the output from a scan must be multiplied by the associated measurement surface area.

The phase mismatch compensation techniques presented in Section 7.2.1 may also be applied to sound power determinations, where appropriate, by replacing the point pressure-intensity index δ_{pI} by the surface-average equivalent $\langle \delta_{pI} \rangle$.[72,73] The cautionary remarks made in that section concerning the accuracy of phase mismatch correction, or compensation, apply equally to sound power determinations, with the added warning that the effect of spatial variation of sign of the normal intensity over a measurement surface increases the random error of a power estimate, by analogy with the effect of frequency variation within a band at a fixed point (see Section 7.3.1): hence it also affects the accuracy of the corresponding estimate of $\langle \delta_{pI} \rangle$.

9.3.3 Random error

Non-systematic errors of sound power determination by discrete point or scanning methods are associated with the following factors: temporal variability of the operating conditions of the source under test or extraneous sources; the presence of air or gas flow; intensity field sampling procedures; sound field characteristics and spectral estimation procedures. This section deals with the last two factors. The variance associated with discrete point measurement is clearly a segment area-weighted sum of the variances of the individual intensity measurements, of which the random errors may be assumed to be statistically independent. The normalised random error of a fixed point p-p intensity estimate is given by eqns (7.21) and (7.22). The corresponding normalised random error of sound power determination is given by

$$e_r(W_p) = \left(\sum_{i=1}^{N} S_i^2 \sigma^2 \{\hat{I}_{ni}\} \right)^{1/2} \bigg/ \left| \sum_{i=1}^{N} S_i I_{ni} \right| \tag{9.18}$$

where ^ indicates 'measured quantity' and p indicates point measurements.

The random error is controlled by the largest area-weighted random errors. If the measurement segments are uniform in area, the normalised random error is given by

$$e_r\{\hat{W}_p(\omega_0, \Delta\omega)\} \approx \frac{\left(\int_{\omega_a}^{\omega_b} \sum_{i}^{N} [(S_{11}(\omega)S_{22}(\omega) - C_{12}^2(\omega) + Q_{12}^2(\omega))/\omega^2] \, d\omega \pi/T \right)^{1/2}}{\left| \int_{\omega_a}^{\omega_b} \sum_{i}^{N} [Q_{12}(\omega)/\omega] \, d\omega \right|} \tag{9.19}$$

The process of continuous movement of the intensity probe across the measurement surface in scanning does not allow the time-average spectral characteristics of a non-uniform sound field at any one point to be reliably estimated, because the time-average characteristics of the field must remain constant during the estimation period. However, on the basis of an assumption that the expression for fixed point random error applies to a set of N contiguous scan segments which are short compared to an acoustic wavelength, and for which the averaging time is sufficiently long to generate independent estimates of intensity, Jacobsen[95] concludes that the normalised error of a sound power estimate based upon scanned samples is inversely proportional to the square root of the *total* integration time for the complete scan in the same manner that it depends upon the total averaging time for all fixed point samples. The resulting expression for normalised random error of estimated sound power is

$$e_r\{\hat{W}_s(\omega_0, \Delta\omega)\} \approx \frac{\left(\int_{\omega_a}^{\omega_b} \frac{1}{N} \sum_{i}^{N} [(S_{11}(\omega)S_{22}(\omega) - C_{12}^2(\omega) + Q_{12}^2(\omega))/\omega^2] \, d\omega/(\pi N T) \right)^{1/2}}{\rho \Delta r |I(\omega_0, \Delta\omega)|} \tag{9.20}$$

The restriction on scan speed corresponding to the assumptions on which eqn (9.20) is based is

$$v/c \ll B/f_0 \tag{9.21}$$

where v denotes scan speed, c is the speed of sound and B is the analysis (or noise) bandwidth of centre frequency f_0.

Several studies have been made of the influence of the scan parameters on the accuracy of sound power determination. Tachibana *et al.*[96] investigated the sensitivity to scanning speed, scan line density and measurement distance of the estimate of sound power radiated by an artificial source comprising two loudspeakers fed with signals of different amplitude and opposite phase which exhibits a strong near field energy circulation pattern. The results indicate little sensitivity to scan speed up to the maximum of $500\,\mathrm{mm\,s^{-1}}$ and that a line density less than or equal to the measurement distance is satisfactory except where the negative partial power passing through the measurement surface is comparable with the positive partial power, in which case both random and bias errors can be large (see Sections 9.3.1 and 9.4.1(b): negative partial power indicator F_+). In a study based on the acquisition of successive short-time estimates of intensity and pressure during scans over surfaces at different distances from industrial sources, Pettersen[97] concludes that a maximum scan speed of $300\,\mathrm{mm\,s^{-1}}$ would, in general, restrict the maximum spatiotemporal sampling error to an acceptable value. In a related study of the normal intensity distribution on a measurement surface parallel to a thin vibrating plate radiating into a reverberant room in the presence of various levels of extraneous noise,[98] Olsen *et al.* conclude that sound power estimates are rather insensitive to scan line density: the results also suggest that values of the field indicator F_{pI} (the difference between surface-average sound pressure and normal intensity levels—see Section 9.4.1(b)) in excess of $10\,\mathrm{dB}$, produced by small measurement distances ($< 20\,\mathrm{cm}$), or the presence of extraneous noise, indicate that slow, dense scanning is advisable.

Unless the normal intensity field has a uniform spectral density within the analysis band, and is also uniform over the entire measurement surface (a highly unlikely condition), the normalised random error will exceed $(BT)^{-1/2}$, where T is the total effective integration time of the measurement. Although eqns (9.19) and (9.20) are similar in form, the spectral quantities involved are derived from signals acquired over largely different regions of the sound field and consequently the resulting error estimates can be totally different in magnitude.

If there are no variations of sign (direction) of normal intensity over a measurement surface, the normalised random error of the sound power estimate is likely to be smaller than that of the constituent intensity estimates. Variations of sign (direction) of normal intensity with position on the measurement surface greatly increase random error in the same way as spectral variations within the analysis bandwidth (see Section 7.3.1). Hence, the presence of strong extraneous noise sources which reverse the direction of the net intensity over portions of a measurement surface will both increase phase mismatch bias error and increase the averaging times to reduce random errors to acceptable values. If the extraneous sources also have a significantly narrower bandwidth than the source, the demand on averaging time is even more severe.

9.3.4   Influence of a diffuse reverberant field

Many determinations of sound power are made in reverberant enclosures such as industrial workspaces, often in the presence of other (extraneous) sources, which must, of course, operate steadily if a sound power determination using intensity is to be successful. As explained in Section 4.8.5, the acoustic field in a reverberant space at frequencies where many acoustic modes are significantly excited by a single frequency source may be modelled as diffuse. It should not, however, be forgotten that a reverberant field excited by a single coherent source is itself globally coherent, whereas the simultaneous action of a number of incoherent sources is to produce correspondingly incoherent components of pressure and particle velocity everywhere in the field. In fact, many sources of practical interest operate at fixed rotational speeds, and the noise generated by speed-dependent source mechanisms is periodic and harmonic, and therefore fully coherent with other sources operating at the same speed. Of course, waves reflected from enclosure boundaries are fully coherent with their source unless mechanisms operating in the propagation medium, such as temperature and velocity fluctuations, produce random scattering. However, if the sources have closely spaced harmonics, or are aperiodic, the phases of the contributions to the sound field at any point from the various sources will be distributed over 2π radians and may be considered to be random: hence the integral of the cross-power spectrum over a frequency band will tend to zero, as therefore will the *correlation* between these contributions. In this case the field may be considered to be ideally diffuse.

Jacobsen[99] derives expressions for the influence of the reverberant diffuse field on the errors of sound power determination. If the reverberant field mean square pressure is not less than the mean square pressure of the direct field of the source averaged over the measurement surface, it increases the bias error due to instrument phase mismatch in accordance with eqn (9.10). Since the direct field pressure decreases with the distance of the measurement surface from the source, this source of error increases in proportion to the measurement surface area. The additional normalised bias error is given by

$$e_\phi(W)_{\text{rev}} = [I_{\text{res}}/(\overline{p^2}/\rho_0 c)_{\text{res}}][(cS_0 T_{60}/13 \cdot 8V)](W_{\text{tot}}/W) \qquad (9.22)$$

where S_0 and T_{60} are the area of the measurement surface and the reverberation time of the enclosure. W_{tot} is the total power radiated by all sources in the enclosure which is not usually known but can be estimated from a knowledge of the acoustic absorption a of the enclosure and the space-average mean square pressure as $W_{\text{tot}} \approx p^2 a/4\rho_0 c$. In order to minimise this source of bias error it is clearly important to minimise the factor $cS_0 T_{60}/13 \cdot 8V = 4S_0/a$.

Although the spatial-average intensity of an ideally diffuse field is zero, the spatial variance is non-zero (see Section 4.8.5). Hence, although the spatial integral over any closed measurement surface of the normal component of the diffuse field intensity is zero, it perturbs the distribution over that surface of the normal intensity of the direct field of the source. The spatial sampling error of a sound power determination based upon a given sampling array is obviously increased if this perturbation increases the spatial irregularity of the field. Analysis of the distributions of sound intensity in the direct field of a source and in its *own* reverberant field indicates that the direct intensity field is dominant over distances of between three and eight times the ideal 'reverberation radius' given by $r = (a/16\pi)^{1/2}$. It may be assumed that perturbation by a superimposed field generated by *extraneous* sources is significant only if

$$W_{tot}/W > a/4S_0 \tag{9.23}$$

On the basis of analysis of a coherent 'diffuse' sound field, Jacobsen derives the following expression for the normalised random sampling error of sound power determination due to the presence of a diffuse field on the measurement surface:

$$e_s(W) = (4S_0/a)(W_{tot}/W)[6N_{eq}(1 + BT_{60}/6\cdot9)]^{-1/2} \tag{9.24}$$

in which N_{eq} represents the equivalent number of statistically independent samples of the intensity field. If $B \gg 6\cdot9/T_{60}$, the error decreases as the square root of the diffuse field bandwidth; as with the bias error, it is advisable to minimise the area of the measurement surface.

The degree of statistical dependence of samples of sound field quantities depends upon the spatial separation of the sample points in terms of an acoustic wavelength and the bandwidth of the estimate. If the field is assumed to be diffuse, the influence of bandwidth is small. Samples taken at separation distances of greater than one wavelength are effectively uncorrelated. Jacobsen presents the following expressions for N_{eq} for discrete point and scanned measurement, respectively:

$$N_{eq}(\text{discrete}) = N/[1 + N(\pi/4)\lambda^2/S_0] \tag{9.25}$$

$$N_{eq}(\text{scan}) = (21/\lambda)/[l + (\pi/2)\, l\lambda/S_0] \tag{9.26}$$

where l is the total length of the scan path and λ is the acoustic wavelength. Equation (9.24) may be considered to give a conservative estimate of the influence of a diffuse field on the sampling error.

The diffuse field generated by extraneous sources also increases the random error of sound power determination associated with spectral estimation, especially if it has a narrow bandwidth. The instantaneous normal intensity on a measurement surface is the product of 'co-operation' between the total instantaneous sound pressure and the total instantaneous particle velocity. These two field quantities comprise the sums of the instantaneous components produced by all sources operating in the fluid (see Section 9.2); hence, the instantaneous intensity has contributions from 'cross terms' between pressure generated by one source and particle velocity generated by another. If the various sources are incoherent, or uncorrelated, these terms will disappear in the time-averaging procedure. However, they complicate the spectral estimation process: for example, if an extraneous source generates noise of much smaller bandwidth than the test source, the averaging time necessary to eliminate its contribution to the local and spatially integrated intensity is determined by its bandwidth and not the source bandwidth.

In an analysis of random error due to the presence of a diffuse reverberant field, Jacobsen[99] makes the assumption that the reverberant field dominates on the measurement surface (see eqn (9.23)) so that cross terms are irrelevant. The additional normalised random error in sound power determination due to a diffuse field generated in an enclosure by an *isolated* source of bandwidth B is given approximately by

$$e_{rev}(W) = (4S_0/a)(6BT)^{-1/2} \tag{9.27}$$

where T is the total effective averaging time of the measurement. (B is the analysis bandwidth if the source bandwidth is greater.)

If the diffuse field is produced by incoherent extraneous sources and $W_{tot}/W > a/4S_0$, the value given by eqn (9.27) must be multiplied by W_{tot}/W. In this case it is the bandwidth of the dominant extraneous sources which is relevant.

It might be concluded from the above expressions that it is advisable to use a measurement surface as close to a source as possible to minimise its area. However, it is well known that errors due to phase mismatch, spatial sampling and spectral estimation increase if the source surface is approached too closely, and a reasonable compromise in reverberant conditions would appear to be to ensure that the measurement surface is sufficiently close to the source for the direct field to dominate, but no less. For example, in a room of $200 \, m^3$ with a reverberation time of $1.0 \, s$, the absorption is $32 \, m^2$, giving a maximum side length for a five-faced cuboidal measurement surface around a small source of $1.25 \, m$. The direct intensity fields of extended sources decrease much more slowly than r^{-2} in the geometric near field and the requirement on S_0 could be relaxed accordingly.

9.4   STANDARDS FOR SOUND POWER DETERMINATION USING SOUND INTENSITY MEASUREMENT

The ISO Standard series 3740–3748 specifies methods for the determination of the sound power of sources in carefully controlled environments by means of measurements of mean square sound pressure at discrete points on a measurement surface which envelops the source. It is impossible to subject many sources of practical concern to these procedures for one or more of the following reasons:

(i) Many sources are too large or heavy to be transported and/or placed in special acoustic test rooms, and they operate in the presence of other sources, so that *in situ* measurement methods cannot be applied;

(ii) Many sources operate as essential components of larger systems, from which they cannot be separated, and of which the noise prevents pressure-based methods from being used.

The primary purpose of a sound power standard is to ensure that any single determination of the sound power of a particular source, made with any measurement system and in any test environment which satisfy the requirements of the standard, lies within the stated tolerance of the true value at a stated confidence level (e.g. 95% confidence of lying within ± 3 dB of the true value). This is a statement of *reproducibility*. Naturally, a standard cannot legislate for variation in operating or installation conditions which alter the sound power output.

Standards for the determination of source sound power using intensity measurements have been developed by both national and international committees of standardisation bodies: a list of current standards will be found in the Appendix. This chapter will be restricted to reviews of the current ISO standards only. Standards normally represent a compromise between the ideal and the practical; they are based largely upon empirical information and experience. This is particularly true of sound power standards based upon intensity measurement. Previously established sound power standards (e.g. the ISO 3740 series) closely specify the test environment and require background noise to be minimised so that confidence ranges for the reproducibility of the result can be reliably specified.

Application of the principle of sound power determination, which suppresses the influence of extraneous sources and of non-ideal acoustic environments, allows source sound power determinations to be made under a wide variety of operational conditions. As a consequence, the sound power standards 9614-1, 9614-2 and ANSI S.12.12 (1992) differ radically from the 3740 series in placing only rather broadly defined constraints on the user in the selection of the measurement surface and the spatial sampling array (or scan path) which are judged to be best suited to the particular geometric form of source and test environment. In order to ensure that the resulting sound power estimate has a sufficiently high probability of lying within the stated range of the true value (which is, of course, never known), these standards require the user to evaluate certain 'indicators' which are used to check certain criteria which form the basis of the judgement of the acceptability of the quality of the instrumentation and of the test environment. These standards also provide guidance on means of improving the test conditions and procedures in cases where these criteria are not satisfied. The diversity of field test conditions precludes the possibility of a 'Precision' (Grade 1) determination, and the ISO standards currently specify only 'Engineering' (2) and 'Survey' (3) grades.

9.4.1 Indicators and their significance

The International Standards 9614-1 and 9614-2 differ fundamentally from their North American counterpart ANSI S.12.12 (1992) in specifying a small number of indicators of which the evaluation is mandatory: the latter Standard specifies a large number of indicators of which evaluation is voluntary in many cases. Only the ISO indicators will be discussed here.

(a) ISO 9614-1: Measurements at discrete points

Four indicators are defined.

F_1: Temporal variability indicator

The principle of sound power determination using intensity measurement is invalidated if the test source, or significant extraneous sources, or the acoustic test conditions, alter during a test. The purpose of the evaluation of F_1 is to check the stationarity (steadiness) of the intensity field by making a series of short-time-average estimates of sound intensity at one point on the selected measurement surface. The shortcoming is that the check is done only immediately prior to, and immediately after, the test—not during it.

F_2: Surface pressure-intensity indicator

This is the difference in decibels between the sound pressure level corresponding to the arithmetic average of the estimates of mean square pressure at the array of measurement points and the sound intensity level corresponding to the arithmetic average of the normal intensities estimated at the measurement points *computed by ignoring the relative signs (directions) of the individual estimates.* It is used to limit the bias error due to instrument phase mismatch. As shown by Figs 6.11, the phase mismatch bias error is no more than ± 1 dB if L_ϕ is greater than 7 dB and no more than $\frac{1}{2}$ dB if L_ϕ is greater than 10 dB. The Standard specifies a bias error factor K for each grade of accuracy, and the dynamic capability of the instrument L_d is defined as $\delta_{pI0} - K$ as shown on Fig. 6.25 (see also Section 6.9.1). F_2 may not exceed the value of L_d corresponding to the grade of accuracy required. (Note: the following indicator F_3 should more correctly be employed for this purpose; see Section 9.3.2, eqn (9.17).)

F_3: Negative partial power indicator

This corresponds to F_2 except that the sound intensity level corresponds to the arithmetic average of estimates *with the sign respected.*

The difference between F_3 and F_2 is a measure of the ratio of partial sound power leaving the measurement surface to that entering it (usually due to the proximity of extraneous sources). Both bias and random errors in the sound power estimate are functions of this ratio (see Sections 9.3.1 and 9.3.3): the maximum bias error is limited by placing a limit of 3 dB on $F_3 - F_2$. This limit also serves to limit the bias error associated with phase mismatch.

F_4: Field non-uniformity indicator

This is equal to the normalised variance (coefficient of variation) of the discrete point samples of normal sound intensity *with sign respected.* It is an indicator

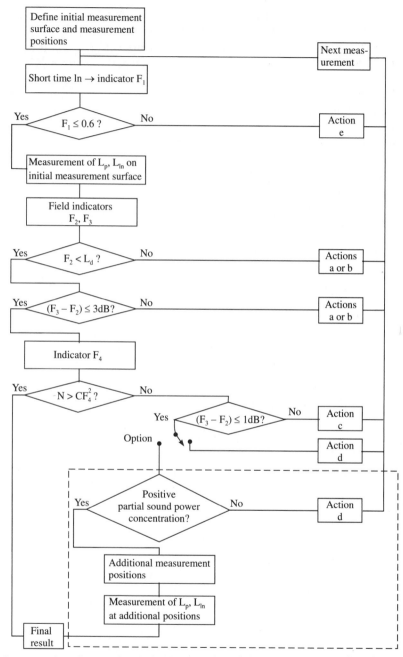

Fig. 9.4. Flow diagram for the implementation of ISO 9614-1. (Extract from International Standard reproduced with permission. Complete copies available from national standards bodies.). *Note*: the path enclosed in broken lines represents an optimal procedure designed to minimise the number of additional measurement positions required on the initial measurement surface.

of the spatial variance of the normal intensity field on the chosen measurement surface. It is used to specify the minimum number of sample points necessary to restrict the uncertainty of the estimates of the spatial mean to within acceptable limits. This application is based upon a fundamental assumption of statistical independence of sample estimates which may often not be justified. F_4 will also be very large if the spatial intensity distribution is strongly bipolar due to the proximity of a strong extraneous source. In such a case it makes much greater sense to evaluate the coefficients of variation of the negative and positive intensities separately.

The ANSI Standard employs a quite different method to qualify the number of sample points: sound power estimates are made using two different arrays of

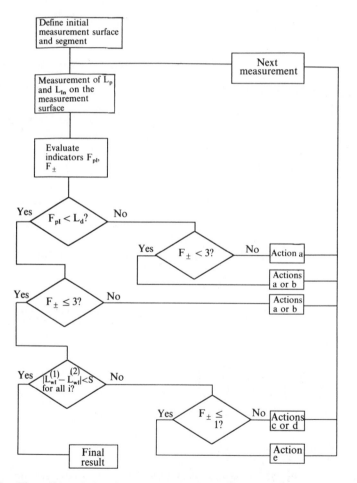

Fig. 9.5. Flow diagram for the implementation of ISO DIS 9614-2 (Extract from Draft International Standard reproduced with permission. *Note*: at the time of writing this has not yet been accepted for publication as an International Standard and therefore has no legal standing.)

sample points and if the difference falls within a given range the number of points is considered adequate. Neither method is entirely satisfactory in principle, but may often prove satisfactory in practice.

A flow diagram for the ISO standard method is presented in Fig. 9.4.

(b) ISO 9614-2: Measurement by scanning

F_{pi}: Sound field pressure-intensity indicator

This indicator corresponds to F_3 in cases where the discrete points are associated with segments of equal area.

$F_{+/-}$: Negative partial power indicator

This indicator serves the same purpose as the limit on $F_3 - F_2$ in ISO 9614-1. They are identical in cases of uniform segment areas.

This Standard also specifies a 'Partial power repeatability check' by which two partial power estimates are made for each segment of the measurement surface using orthogonally oriented scan patterns: it is required that the difference is smaller than a specified value. This serves to monitor the stationarity of the intensity field.

A flow diagram for this standard method is presented in Fig. 9.5.

9.4.2 Ameliorative actions

In cases where the field conditions and/or the chosen measurement surface do not satisfy the criteria specified in the standards, the user is advised to take certain ameliorative actions. These consist either in moving the measurement surface (normally towards the source surface) or installing temporary screening. The latter strategy is especially useful where very strong extraneous sources greatly distort the normal intensity distribution on the measurement surface. The erection of a simple wooden or gypsum board screen is usually sufficient: it alters the source power in principle, but in practice it has little influence except at low frequencies (see Section 9.5.3), especially if it is covered in a layer of efficient sound-absorbent material.

9.5 EXAMPLES OF SOURCE SOUND POWER DETERMINATION

Innumerable determinations of sound power have been made using sound intensity measurement since commercial systems became available around 1980. The examples described in the following sections have been selected because they demonstrate features of particular practical interest or significance.

9.5.1 Steam turbine generator unit

The sound power of a 660 MW steam turbine generator unit together with that of its ancillary equipment was measured in a reverberant turbine hall ($T_{60} = 3\,\text{s}$) containing two such units and measuring $58 \times 144 \times 31$ m.[100] Many other sources of extraneous noise were operating simultaneously. Two features of this investigation are of particular interest: the generator sound power level is nearly 10 dB(A) less than the pilot exciter and it has such a large surface area that its surface normal intensity levels are extremely low compared with the ambient sound pressure levels; and measurements were made of *pu* coherence in an attempt to classify the form and identify the origins of the sound field in various regions close to the sources. As described in Section 8.4, *pu* coherence is normally high in source near fields, very low in fields generated by many incoherent sources, and is unity, but subject to analysis bias error, in reverberant fields generated by a single (coherent) source.

The layout of the plant, together with some sound intensity component levels (dB(A)) at discrete points, is shown in Fig. 9.6. Intensity scans were made at a maximum speed of $250\,\text{mm s}^{-1}$ at a distance of 10 cm from the surface of the generator with a scan line separation of 120 mm. Measurements of *pu* coherence were made at the points marked X. The overall dB(A) sound power levels of the four main components are also indicated in the figure. A-weighted narrow band sound power level spectra and the equivalent 1/3 octave band spectra are shown in Figs 9.7 and 9.8 respectively. The spectra of F_{pl} and $F_{+/-}$ are presented in Figs 9.9 and 9.10. The field indicator for the generator measurement surface is comparable with the pressure-residual intensity index of the instrumentation (22 dB) over much of the frequency range; the associated phase mismatch bias error is responsible for much of the negative intensity content of the narrow band spectrum of generator sound power. Correction for this error made according to Section 9.3.2 reduces the

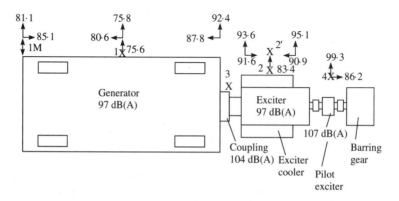

Fig. 9.6. Sound intensity distribution around a generator unit, dB(A) sound powers and measurement locations of sound pressure/particle velocity coherence (denoted by X).[100]

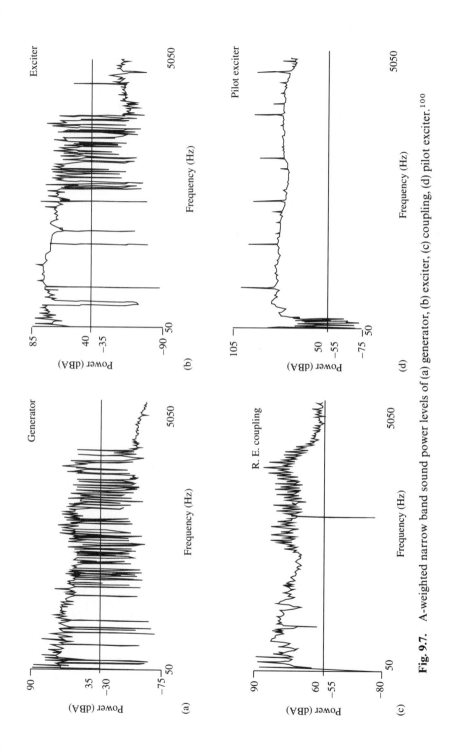

Fig. 9.7. A-weighted narrow band sound power levels of (a) generator, (b) exciter, (c) coupling, (d) pilot exciter.[100]

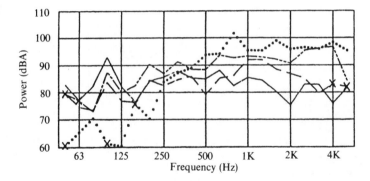

Fig. 9.8. A-weighted 1/3 octave band sound power levels of the generator system: —— generator; .... pilot exciter; ––– exciter; – – – – coupling.[100]

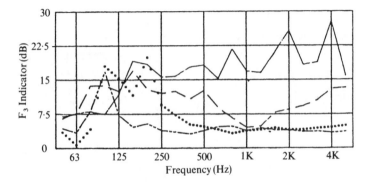

Fig. 9.9. One-third octave band values of indicator F_3: —— generator; .... pilot exciter; ––– exciter; – – – – coupling.[100]

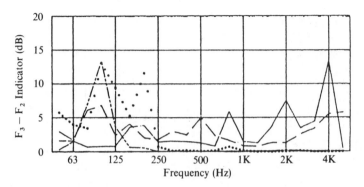

Fig. 9.10. One-third octave band values of indicator $F_3 - F_2$: —— generator; .... pilot exciter; ––– exciter; – – – – coupling.[100]

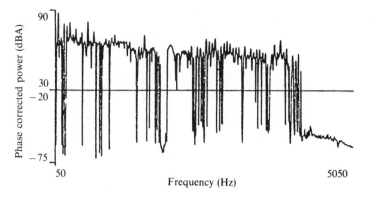

Fig. 9.11. A-weighted phase-mismatch-corrected narrow band sound power levels of the generator.[100]

proportion of negative power as shown by Fig. 9.11, but there must remain some doubt about the significance of the many negative 'spikes' in the spectrum, which could possibly have been resolved by (costly) repeatability checks.

The *pu* coherence close to the generator is low and does not increase with fineness of frequency resolution, thereby indicating that resolution bias error is not responsible and that the local field approximates to a diffuse field comprising contributions from many incoherent sources. The increase of coherence with distance from the exciter reveals the increasing influence of the strong direct field from the pilot exciter. Close to the rear end generator coupling the coherence is increased from 0·5 to unity by refining the frequency resolution from 6·25 to 1·25 Hz, clearly indicating that the field is reverberant and generated by one main source. The coherence spectrum beside the pilot exciter is characteristic of a direct field which dominates lesser reflected components.

9.5.2   Hydroelectric installation

A principal element of any measurement standard is a statement of confidence of reproducibility. Reproducibility necessarily implies the application of a standard by different users in different locations. As part of a programme of development of the Norwegian Standard for sound power determination based upon scanned intensity measurements (NS-INSTA 121) tests were carried out by a Nordic group on a hydroelectric installation.[101] Four teams having varying degrees of experience of intensity measurement and using various combinations of commercial equipment, applied the draft Standard to a turbine and a compressor in a very reverberant hall: Figs 9.12(a–d) show the results. It is not possible to establish whether the observed differ-

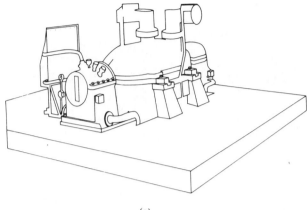

(a)

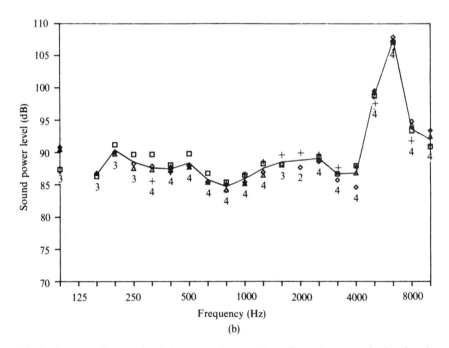

(b)

Fig. 9.12. Sound power level determinations made by four teams on a hydroelectric installation: (a) turbine; (b) measured turbine sound power level spectra. □, Team 1; +, Team 2; ◇, Team 3; △, Team 4.[101]

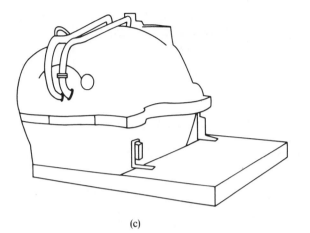

(c)

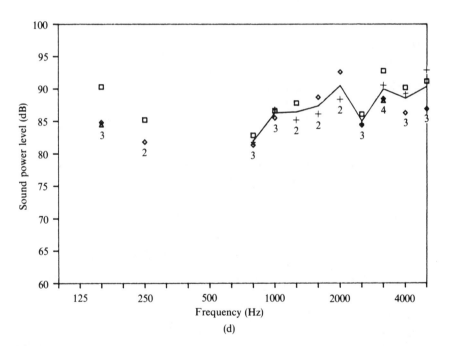

(d)

Fig. 9.12—*contd.* (c) Axial compressor; (d) measured compressor sound power level spectra (results omitted if $F_2 > 10\,dB$); □, Team 1; +, Team 2; ◇, Team 3; △, Team 4.

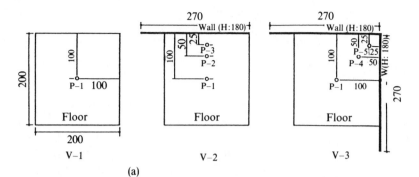

(a)

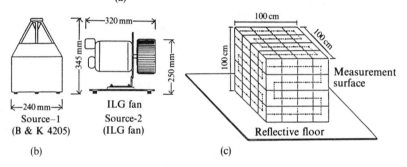

(b)   (c)

Fig. 9.13.   (a) Three test configurations; (b) reference sound sources; (c) cubic measurement surface.[102]

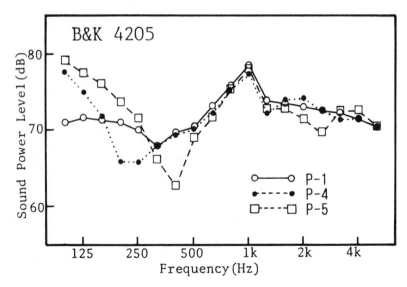

Fig. 9.14.   Source sound power levels of source-1 measured under test condition V-3.[102]

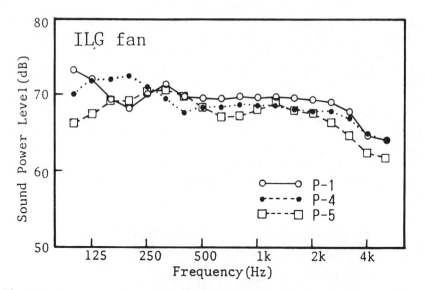

Fig. 9.15. Source sound power levels of source-2 measured under test condition V-3.[102]

ences are statistically significant, in the absence of repeatability data for each team.

9.5.3 Influence of environment on source sound power

The influence of the test environment on the sound powers of two forms of reference sound source was determined by discrete point sound intensity measurement.[102] The measurement arrangements are shown in Fig. 9.13: the effects of the proximity of reflective surfaces on the sound power of an electroacoustic source and an aerodynamic source are illustrated by Figs 9.14 and 9.15 respectively. It is clear that the close proximity of reflective surfaces greatly influences 1/3 octave band sound power, especially at the lower frequencies: of course, it would influence tonal power even more strongly, since the reflected field is fully correlated, unlike finite frequency band reflections. It was found that manually scanned measurements agreed well with the discrete point results.

Sound intensity measurements were also used to evaluate the validity of the 'Waterhouse correction' in a study of the determination of sound power in reverberation rooms at low frequencies:[102] overall the correction got a 'clean bill of health'.

9.5.4 Gated sound power measurement

Many sources of noise operate according to a repeated cycle; sound radiation is often dominated by different mechanisms (or events) during different parts of the cycle with which it is possible to associate the measured sound intensity field and the radiated sound power for the purpose of source identification (see also Section 10.1.2). Of course, allowance should be made for time of flight from the source surface to the measuring point and the influence of the 'vibrational memory' of a structure in response to an applied disturbance.

Examples of the results of the application of this technique to the noise radiated by a garden tractor, together with a schematic of the equipment, are shown in Figs 9.16–9.18.[103] Repeatability and consistency are seen to be good (Table 9.2). The extraction of the exhaust sound power spectrum allows the principal peak to be unequivocally attributed to the eighth to twelfth har-

Fig. 9.16.   Garden tractor used for tests of gated sound power.[103]

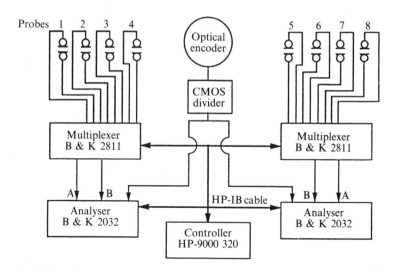

Fig. 9.17. Schematic of equipment used to measure gated sound power.[103]

Table 9.2 Comparisons of ungated and sum of component gated sound powers[103]

	Gated sound power (mW)			Sum of gated sound power	Ungated sound power (mW)
Combustion	*Exhaust*	*Intake*	*Compression*		
1·64	5·03	2·64	1·96	11·27	11·71
1·61	4·91	2·56	1·88	10·96	10·95
1·64	4·67	2·61	1·95	10·87	11·04
1·56	5·02	2·58	1·97	11·13	11·18
Corresponding sound power data in dB(A)					
92·1	97·0	94·2	92·9	100·5	100·7
92·1	96·9	94·1	92·7	100·4	100·4
92·1	96·7	94·2	92·9	100·4	100·4
91·9	97·0	94·1	92·9	100·5	100·5

monics of the engine firing frequency, which presumably reflects the performance of the silencer. The relatively coarse frequency resolution of the spectral plots (108 Hz) is a result of the need to use a transient window on the raw data.

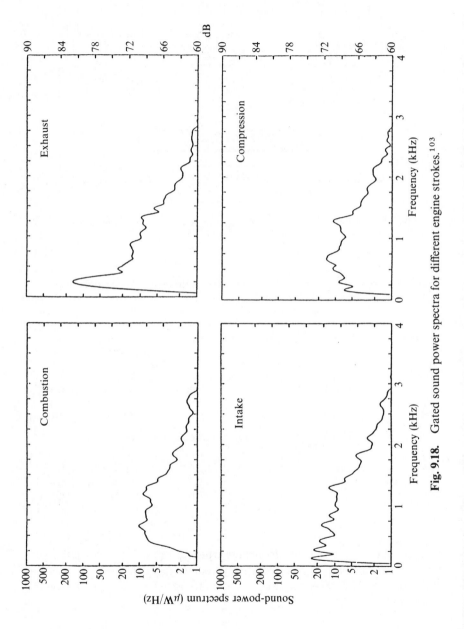

Fig. 9.18. Gated sound power spectra for different engine strokes.[103]

CHAPTER TEN

Other applications of sound intensity measurement

10.1 SOURCE LOCATION AND CHARACTERISATION

10.1.1 Principles

It is a generally accepted principle that 'control at source' is the most desirable form of noise control measure. Indeed, the European Community directives on machinery noise control specifically exhort those responsible for their implementation to make this target a priority. Airborne noise is often the end product of a chain of events and mechanisms of which the origin is some mechanical or fluid process such as gear meshing or combustion. Sound intensity measurement is most frequently applied at the point in the chain where solid surface vibration generates sound waves in a contiguous fluid (usually air), or where airborne sound which is generated within a system emerges into the open, as in ventilation systems. In this context we may restrict our definition of the 'source' which we hope to identify to actions occurring at this interface. We may also wish to study the structure of a sound intensity field as an indicator of the form of the mechanism creating the sound, for the purpose of source characterisation.

The problem of identifying a source has been addressed qualitatively in Chapter 8 where it is emphasised that 'pressure' sources and 'power' sources are not generally one and the same, and that in many cases of practical concern there is no simple relationship between sound pressure at any specific point in a field and the measured sound intensity or sound power. However, in many noise problems it is not the sound pressure at any single point which is of concern but the impact of the noise over a large region such as a residential area close to an industrial plant or the workspace of a factory. In such cases, reduction of the sound power generated by a source is the primary target of the noise control engineer.

Two techniques are commonly used for surface source investigation by means of sound intensity measurement: one quantifies the spatial distribution of power radiated by a surface by means of measurements close to the source surface; the other is based on the projection of intensity vectors measured in the geometric near field back to the source surface. In the former, a measurement surface is selected to be approximately conformal with the physical surface under investigation and at an average distance from it of between 50 and 100 mm. The surface is divided into segments, often associated with individual mechanical components, and the partial sound powers passing through each surface are determined by either scanned or fixed point normal intensity measurements. It is implicitly assumed that each measured partial power is generated by the region of the source surface associated with the corresponding segment of the measurement surface; this will not be so where surface sources radiate strongly in directions far from normal to the surface. It has become common practice to generate intensity contour maps by means of spatial interpolation of partial power data and to superimpose them onto graphical representations of the system under investigation. Reference to eqn (4.11b) will reveal that such a procedure based upon a two-dimensional slice through an intensity field can produce phantom 'sources' and quite misleading conclusions.

As noted in Section 8.3.1, the complexities of the hydrodynamic near fields of vibrating structures adversely affect the utility of close field intensity and partial power measurements for the purposes of source location and quantification, and random errors are exacerbated by reversals of direction of energy flow along a scan path (Section 7.3). Near field circulation is exhibited most strongly by inefficient radiators, in particular by panel structures vibrating at frequencies well below their critical frequencies. For example, intensity measurements made within about 100 mm of window panes subjected to traffic noise are often found to be unreliable, particularly since the sound energy is transmitted principally by non-resonant vibration, whereas the near field is generated mainly by resonant vibration of the relatively lightly damped structure.[33] The data from such measurements are of little use for the purpose of identifying and quantifying 'the source'. In cases where it is suspected that strong near field circulatory energy flow is present it is advisable to make slow sweeps over proposed measurement surfaces with an appropriate setting on the instrument (e.g. instantaneous or exponentially averaged spectra) to establish the extent of the near field in a direction normal to the surface, so that it may be avoided.

Nevertheless, near field estimates of partial power distribution, which generate 'rank order' information, may often be successfully used to investigate machinery noise, particularly that of internal combustion engines. How can this be, given the difficulties presented by near fields? There are two basic reasons: first, engine blocks, crankcases and many industrial machinery and plant components are relatively stiff structures which radiate efficiently, producing only weak near fields; second, the excitation mechanisms and resulting surface vibrations have broad band frequency spectra. The acoustic near field of a vibrating surface is formed by interference between the acoustic disturbances generated by the accelerations of the various elements of the surface. The mean intensity at a point in a time-stationary sound field is equal to the zero-time-delay cross-correlation of sound pressure and particle velocity (eqn (5.9)). The mean intensity within a finite frequency band is equal to the integral over the band of the real part of the cross-spectral density of these quantities (eqns (5.11) and (5.13)), which may be positive or negative: this integral corresponds to a band-limited zero-time-delay cross-correlation. If the autospectrum is rather uniform over the band, and the phases of the cross-spectral components are distributed rather uniformly over the range $0-2\pi$, the resulting correlation function will tend to zero. When a surface vibrates at many frequencies within a given band, the vibrational acceleration of any two points tends to lose correlation as the distance between the points increases, even if the vibration is perfectly coherent over the whole surface at any one frequency: this is because the spatial variations of phase with distance differ from frequency to frequency. The region of correlated motion thus tends to contract as the bandwidth increases. The result is that contributions to the *pu* cross-spectrum at any field point of pressure generated by one element of the surface and particle velocity generated by another element of the surface, tend to cancel when integrated over the vibrating surface. Hence the near field interference patterns and their associated convoluted circuits are 'smeared out', rather like those in a duct (Figs 4.15 and 4.16).

The second technique for locating, but not quantifying, principal sources of radiated power is applied in the geometric far field. If a particular region of a sound-generating system dominates the radiated field, a two- or three-dimensional map of the intensity vector in the geometric near field (where the source subtends a large solid angle at the observer position) will usually indicate the location of the principal source *via* backward projection of the intensity vectors to the source surface. Examples of the results of this technique are presented later in this section. This technique is unreliable if applied to narrow band sound fields within reflective enclosures because of curvature of the energy flux streamlines, but works well with broad band sources in free field or in large enclosures.

In cases where it is wished to apply source location techniques to a source of interest which operates in the presence of strong extraneous sources which contribute significantly to the net intensity field in the vicinity of the item of interest, it is advisable to employ a temporary screen with an absorbent surface to reduce the influence of the 'parasitic' source(s). It is particularly difficult to interpret the intensity distribution on a measurement surface which lies between two sources of rather similar power (e.g. between a motor and a gearbox). The temporary screen technique can be applied here provided that the sound-absorbent surface is efficient in the frequency range of interest; otherwise it will reduce the net intensity to immeasurably small values. Alternatively, a rigid reflective screen may be employed with the attendant risk that its presence may alter the sound powers radiated by sources in its proximity.

Since reactive intensity is strong in the near field of idealised point sources, it is tempting to believe that the presence of a strong reactive intensity component near a physical object is an indication of the presence of a source. This is often, but not necessarily, so. For example, interference between a wave

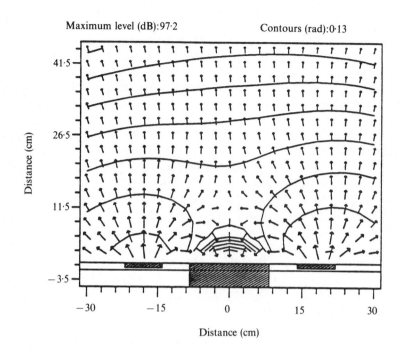

Fig. 10.1. Active intensity and wavefronts at 200 Hz measured in a plane perpendicular to a rigid baffle containing a tube at resonance flanked by two in-phase piston sources.[105]

propagating along a surface from a nearby source and the near field produced by vibration of the surface can produce reactive intensity fields close to the surface which closely resemble those produced by strong local sources, even though these do not, in fact, exist. Unfortunately, reactive intensity is not always a very helpful indicator, because it tends to be strongest in the vicinity of weakly radiating vibrating surfaces. It is also strong in pure-tone reverberant fields, as seen in Chapter 4, in which regions of concentration of reactive intensity in no way indicate the presence of active sources as revealed by the *divergence* of the reactive intensity vectors. The distribution of reactive intensity near a vibrating surface corresponds closely to the distribution of the normal velocity field at low frequencies where the near field is strong:[89,104] but the phase of the motion, which is critical in controlling radiated sound power, is not revealed and therefore it is not as useful as direct vibration measurement by accelerometer or laser. It is possible to combine reactive intensity plots with pressure iso-phase contours which provides a revealing picture of the nature of the sound field. Figures 10.1–10.4 show the fields near to the mouth of a tube flanked by two small piston sources at frequencies corresponding to the resonance and anti-resonance frequencies of the tube.[105]

Reactive intensity

Maximum level (dB):110

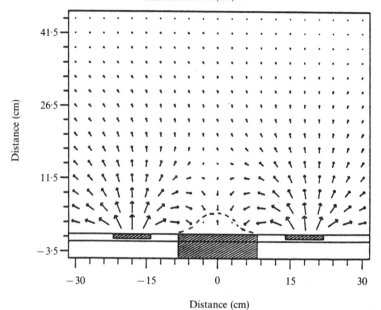

Distance (cm)

Fig. 10.2.   Reactive intensity at 200 Hz measured in a plane perpendicular to a rigid baffle containing a tube at resonance flanked by two in-phase piston sources.[105]

10.1.2   Practical examples

The following section presents a selection of examples of source location and quantification. Results of the application of the technique of near field partial power ranking on automotive engines are presented in Fig. 10.5 and Table 10.1.[106] A good example of the application of sound intensity measurement to the identification, quantification and rank ordering of noise sources in an industrial compressor plant installation, together with an analysis of the implications for the selection of noise control measures, is provided by Johns and Porter.[107] The use of the vector projection technique to locate major source regions is illustrated in Figs 10.6–10.10.

Initial surveys of a 'new' source are most quickly and conveniently performed in dB(A). The probe is swept over the entire surface of the source system at a distance of between 50 and 100 mm, with the axis of the probe directed towards the source surface; regions of high positive (outgoing) intensity are noted, together with any regions of negative intensity. Then the probe axis is reoriented to lie approximately parallel to the source surface and the scan is repeated. This latter procedure will reveal the presence of any highly localised, normally directed power flow, such as that through a leak or weak

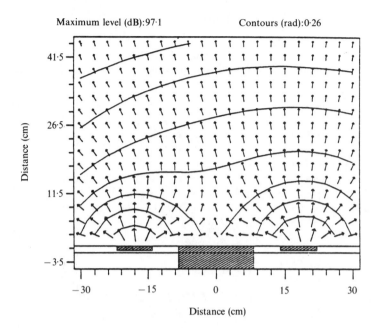

Fig. 10.3.　Active intensity and wavefronts at 400 Hz measured in a plane perpendicular to a rigid baffle containing a tube at anti-resonance flanked by two in-phase piston sources.[105]

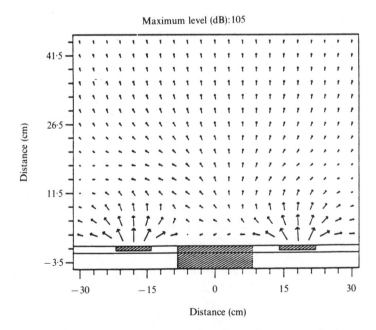

Fig. 10.4.　Reactive intensity at 400 Hz measured in a plane perpendicular to a rigid baffle containing a tube at anti-resonance flanked by two in-phase piston sources.[105]

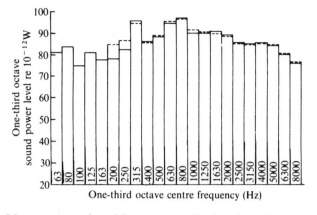

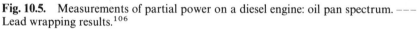

Fig. 10.5. Measurements of partial power on a diesel engine: oil pan spectrum. – – – Lead wrapping results.[106]

Table 10.1 Comparison of overall sound power levels obtained by the acoustic intensity, surface intensity and lead-wrapping methods

Engine part	Overall sound power level Re. 10^{-12} Watts at 1500 rpm and 542 N m		
	Acoustic intensity	Surface intensity	Lead-wrapping
Oil pan	102·7	103·3	102·6
Exhaust manifold, turbocharger, cylinder head and valve covers	101·4	—	101·6
Aftercooler	100·8	101·9	100·6
Engine front	95·0	—	100·0
Oil filter and cooler	91·1	93·4	98·1
Left block wall	97·4	94·6	97·3
Right block wall	94·8	93·3	97·3
Fuel and oil pumps	91·5	—	96·3
Sum of the parts	108·1	—	108·8
Sum of oil pan, aftercooler, oil filter and cooler, and block walls	106·1	106·4	106·7
Bare engine sound power	108·1[a]	—	109·5

[a] Calculated from sum of parts.

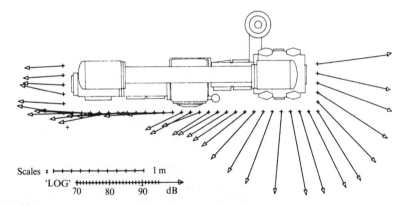

Fig. 10.6. Distribution of intensity around a compressor unit in the 200 Hz 1/3 octave band (courtesy of CETIM), Senlis, France).

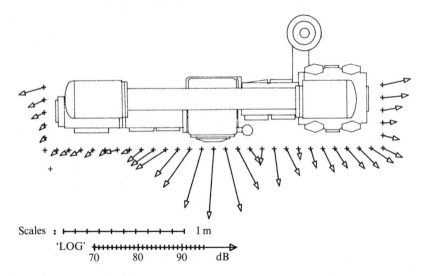

Fig. 10.7. Distribution of intensity around a compressor unit in the 800 Hz 1/3 octave band (courtesy of CETIM, Senlis, France).

area in the source surface, as well as any power flow due to the near field circulation phenomenon; the indication will take the form of intensities of suddenly varying sign. (It is appropriate to mention here that, in my experience, the human auditory system is also quite effective in locating medium to high frequency source regions. If the ears are fully cupped by hands formed into quarter 'spheres' which are used to push the pinnae into positions approximately at right angles to the side of the head, sources may often be accurately pin-pointed, even in very noisy, reverberant conditions: outdoor sources are even more effectively revealed. It is surprising how many acousticians fail to take advantage of this free, portable facility. Readers may also

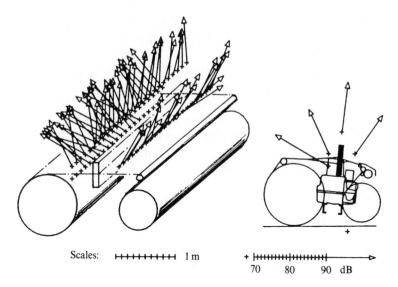

Scales: ├──────┤ 1 m   + ├────┼────┤ ──▶
70   80   90 dB

Fig. 10.8. Mean intensity vectors in the field of a loom in the 1 kHz 1/3 octave band.[108]

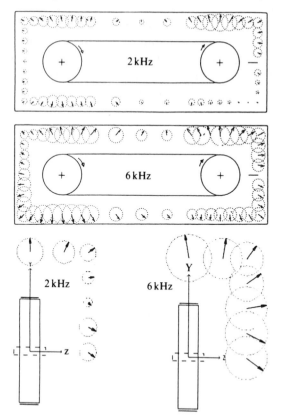

Fig. 10.9. Mean intensity vectors in the field of a synchronous belt system.[109]

like to try this expedient while listening to their domestic audio system—it makes a £200 system sound like a £1000 system—unfortunately the position is tiring, and somewhat anti-social.)

Sound intensity probes are extremely effective in locating leaks in partition structures, and they should be routinely used to check out installations in transmission suites prior to transmission loss measurement. They are also very useful in proving built-up dry constructions, such as demountable office partitions and music booths, machinery enclosures and folding partitions.

A variation on continuous intensity measurement has been used to investigate cyclic sources, in which the probe signal sampling is synchronised with the cycle, and time windowed (gated) so as to reveal more clearly the relationship between the various source mechanisms and the radiated sound. It was originally applied to tyre noise measurement[111] and more recently to engines.[112] An example of the type of results obtained is shown in Fig. 10.11. It should be noted that there exists no unique time delay between *extended* source action and wave reception: also, structural ringing can extend over a significant part of a cycle.

In an attempt to distinguish between various uncorrelated sources of sound power generation and transmission, a 'selective intensity' technique has been under development for some time. The principle is based upon the possibility of judiciously choosing and positioning a 'reference' transducer so that its signal is strongly representative of the kinematic or dynamic behaviour of some local region of a complex, extended source/transmission system. Then the cross-spectra between the signals from the intensity probe and the reference transducer are separately computed, before combination to form the

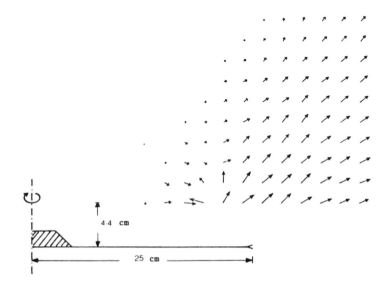

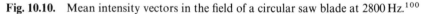

Fig. 10.10.   Mean intensity vectors in the field of a circular saw blade at 2800 Hz.[100]

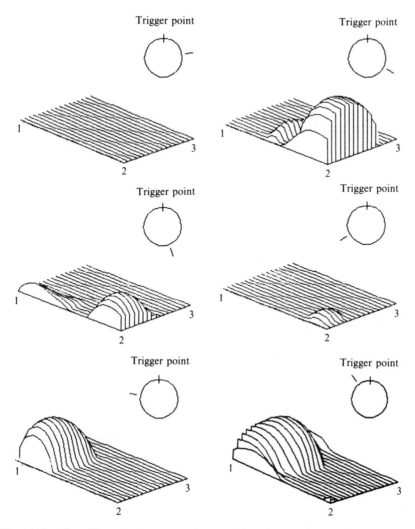

Fig. 10.11. Gated intensity maps on a four-speed medium diesel at various points in a cycle.[112]

coherent cross-spectral component of the two probe signals. By this procedure, those components of the intensity probe signals which are not statistically related to the selected source region are suppressed. The original procedure[113] was based upon partial coherence techniques, and required more than two signals to be analysed simultaneously; a later development[114] simplified the procedure so that conventional two-channel FFT analysers could be used. In a recent development,[115] the selective technique has been combined with the intensity scanning procedure, in order to minimise test time. Figure 10.12 shows some near field intensity distributions measured on

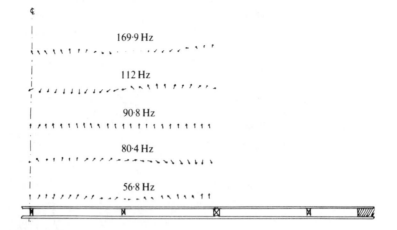

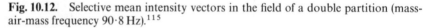

Fig. 10.12. Selective mean intensity vectors in the field of a double partition (mass-air-mass frequency 90·8 Hz).[115]

a lightweight double wall panel, using the selective swept intensity technique; the efficient radiation at the mass–air–mass resonance of the structure is clearly revealed.

Attempts to use sound intensity measurement to locate 'sources' of low frequency (<200 Hz) sound on the boundaries of land vehicle cabins have been conspicuous by their lack of success. Given the combination of periodic excitation, highly reactive field conditions and thin vibrating structural boundaries, this should not come as a surprise. Even if a distribution of sound powers can be attributed to the enclosure boundaries by means of near field sound intensity measurement, changes to any part of that boundary, or to the acoustic conditions in the enclosed volume, can drastically alter that distribution. More seriously, the accompanying changes in the sound pressure level at specific positions within the enclosure are not related to changes in sound power in any systematic manner. It must be concluded that sound power is not a useful design or modification criterion under these conditions. This conclusion is supported by the results of a theoretical analysis of the effect of installing a floor in an idealised aircraft cabin:[116] at frequencies above about half the ring frequency of the structural cylinder, the acoustic pressure distributions were greatly altered by the floor, but the wall surface intensity distributions and radiated power were not significantly affected.

Evidence of the transformation of surface source regions into 'sink' regions by alteration of a different surface radiating into the same space is found in Ref. 117. Various forms of noise control treatment were applied to the major surfaces of the aft galley area of a production airliner. Measurement of the partial sound powers radiated by these surfaces before and after treatment showed that the aft surface was converted from a source into a sink, and the sound power radiated by the galley ceiling actually increased. Fortunately the changes did result in reducing the sound pressure level in the galley. There is no way in which the changes observed could have been predicted theoretically, depending as they do on subtle changes of relative phase distributions of the pressures and normal velocities on the radiating surfaces.

In order further to investigate this problem, a simple, idealised, two-dimensional model of a car interior has been analysed. It consists of a rectangular box measuring $1 \times 2\,$m; the two long sides are assumed to be rigid, one end has a locally reacting acoustic impedance which represents a carpet-like finish, and the other boundary comprises an elastic panel which is excited either by a point force or by a plane incident sound field. Intensity vector and sound pressure levels have been computed for a wide range of excitation conditions and frequencies from which two examples have been selected.

In Fig. 10.13(a), the panel is excited by a plane sound wave at an angle of incidence of 45°. The frequency of excitation is 110 Hz which does not correspond to a structural or acoustic resonance. The figure shows the effect of suppressing the vibration of the first panel mode, which hardly alters the total vibration level because this mode is being excited well above its resonance frequency of 26 Hz; the changes in radiated power (21·4 dB), intensity and sound pressure level distribution are, however, considerable, with no discernible relationship between them.

In Fig. 10.13(b) the panel is excited at resonance in its second mode (105 Hz) by a point force. The figure shows the effect of increasing the panel damping by a factor of ten. Again the changes in radiated sound power and sound pressure level do not appear to be strongly related. This highly idealised example may exhibit more extreme behaviour than the physical reality, but it serves as a warning to those who might otherwise seek to use sound intensity measurement as a diagnostic tool in circumstances in which it is not appropriate.

As a counterbalance to these gloomy prognostications, Figs 10.14(a) and (b) show some intensity distributions measured in a bus which clearly indicate the back axle as a primary source of noise.[118]

Surveys of the intensity distributions in the near fields of underwater vibrating structures have also been carried out with a view to identifying and suppressing sources of sound power. The principles of measurement are exactly the same as those described in Chapter 5 for airborne sound. A pair of phase-matched hydrophones are separated by a distance greater than that used in air in proportion to the ratio of speeds of sound—a factor of approximately 4·2. Care has to be taken that the sound field does not significantly vibrate the hydrophone assembly. Because the critical frequencies of plates vibrating in water are about nineteen times those in air, sound radiation efficiency is generally extremely low; near fields are very strong, near field circulation of energy into and out of the vibrating structures is strongly evident, and interpretation of near field intensity measurements in terms of far field effects is difficult. Some examples of such measurements are shown in Figs 10.15(a) and (b).[119]

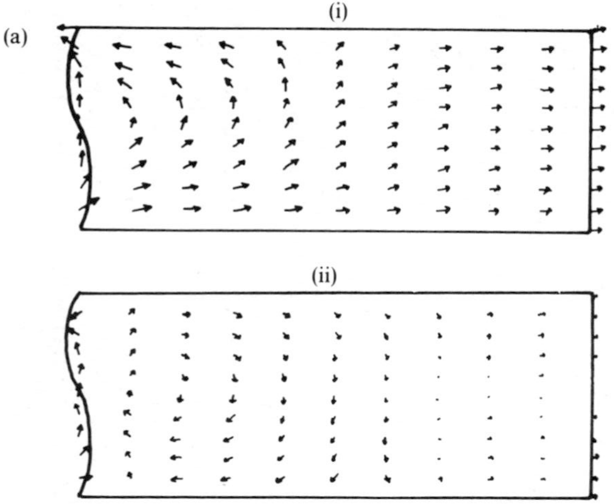

0	-2	-5	-11	-18	-14	-5	7	15	21	23
0	-3	-7	-14	-14	-10	2	7	16	22	24
-1	-3	-20	-15	-5	-3	-6	9	19	26	28
-3	-11	-10	-6	6	7	1	13	26	36	42
23	48	34	32	31	32	31	31	31	31	32
2	5	8	11	13	13	10	7	18	23	23
1	3	4	7	9	9	6	-2	14	19	20
1	2	4	5	7	7	12	-7	10	16	18
1	1	3	5	6	7	4	-10	9	6	17

Reduction in sound pressure level (dB)

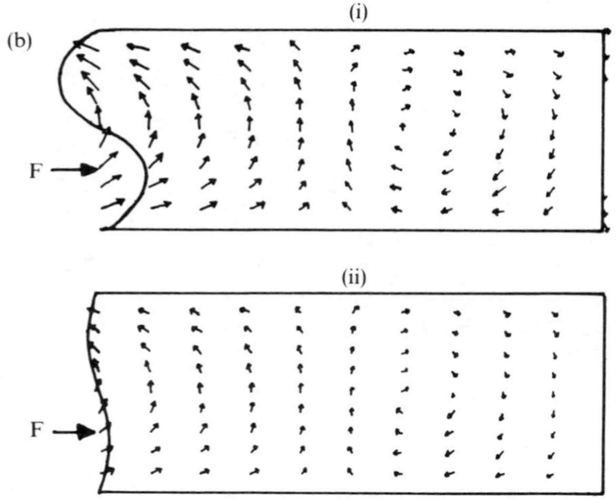

21	21	21	20	20	20	20	21	15	11	9
21	21	21	21	20	19	20	19	13	9	8
21	21	21	20	20	20	20	17	12	7	6
21	21	20	19	17	17	18	15	7	3	3
0	0	0	0	0	0	0	0	0	0	0
21	20	19	18	17	17	22	17	7	3	2
21	20	21	20	19	20	21	20	12	7	6
20	21	20	20	20	20	21	21	15	10	8
20	21	21	20	20	21	22	22	15	11	9

Reduction in sound pressure level (dB)

Fig. 10.13. Changes in mean intensity distributions and sound pressure level reductions in an idealised vehicle passenger compartment: (a) 110 Hz—off-resonance—removal of fundamental panel mode (natural frequency—26 Hz, (i) $L_w = 36{\cdot}5$ dB, (ii) $L_w = 15{\cdot}1$ dB; (b) 105 Hz—resonance of panel mode—increase of panel damping by a factor of 10, (i) $L_w = 41{\cdot}6$ dB, (ii) $L_w = 34{\cdot}5$ dB.

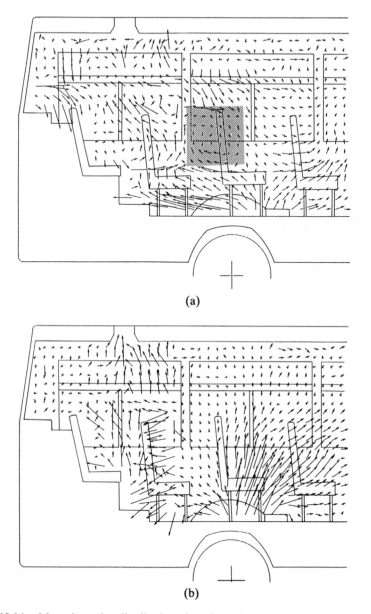

(a)

(b)

Fig. 10.14. Mean intensity distributions in a bus on a chassis dynamometer at a simulated speed of 60 km/h: (a) 100 Hz 1/3 octave band; (b) 250 Hz 1/3 octave band.[118]

For those who attempt to use sound intensity measurement as a source location and diagnosis technique, the measurement process is a means to an end; that end is usually the reduction of radiated noise. It will have become clear from the preceding strictures that near field intensity distributions do not

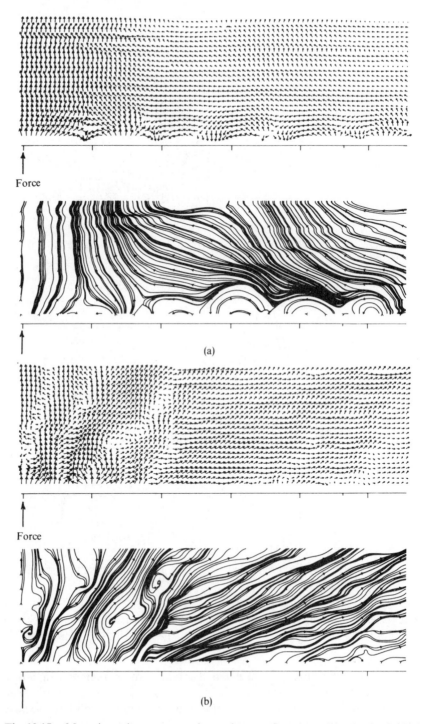

Force

(a)

Force

(b)

Fig. 10.15. Mean intensity vectors and sound power flux streamlines in the field of a point-force excited cylinder submerged in water: (a) 2 kHz; (b) 16 kHz.[119]

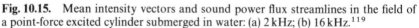

offer an unambiguous guide to the engineer who is seeking to specify physical measures to reduce noise. However, when used in conjunction with other forms of experimental and theoretical analysis, such as modal analysis, transfer function measurement and finite element analysis, it offers a powerful tool for probing the nature and causes of sound fields generated by complex mechanical systems.

10.2 MEASUREMENT OF SOUND ABSORPTION

The sound power absorbed by insonified material bodies may be measured in the same way as radiated power by enclosing them in a notional surface and determining the power transported through the surface from measurements of normal surface intensity. Provided that a body absorbs more than about 25% of the sound power incident upon it, this measurement can be performed quite simply. If the body is less absorbent, significant measurement errors can occur.

It is not so simple to determine the absorption coefficient of a surface because it is not generally possible to measure the *incident* intensity alone; it is always the net (incident minus reflected) intensity which is measured. Incident power has, therefore, to be inferred from measurements of sound pressure, which requires the assumption of an idealised incident wave field (e.g. plane progressive, diffuse). In spite of this limitation, sound intensity measurement offers one important facility not available with conventional absorption measurement techniques, namely, the possibility of evaluating the *in situ* absorption of an individual component of a multi-element system, such as an absorbent panel in a recording studio. This capability can prove extremely valuable in performance evaluation.

Intensity may be applied to isolated specimens of absorbing material, installed in a normal reverberation chamber, according to the ISO standard method. The absorbed power may be determined by enclosing the specimen within a measurement surface and employing either the fixed point or scanning techniques to estimate the surface integral of the normal intensity. The *incident* power cannot be directly measured; it can only be inferred from the space-average mean square pressure in the reverberant field according to the hemi-diffuse field relationship $I = \langle p^2 \rangle / 4\rho_0 c$. This approximation is largely responsible for the considerable discrepancies observed between the measurements of sound absorption coefficients of samples of the same material published by different laboratories, and little reliable evidence for the influence of volume-distributed scatterers (acoustic clouds) on the accuracy of this relationship appears to be available. Unfortunately, therefore, the application of intensity measurement does not help in this respect.

The essential problem with such a measurement of absorption coefficient is that the pressure-intensity index δ_{pI} near the surface of a specimen in a reverberation chamber is likely to be rather large, especially when the absorption coefficient is low, and relatively small errors in measured absorbed sound power level correspond to large fractional errors in derived absorption coefficient. The space-average sound pressure level at the rigid wall of a reverberation chamber is 3 dB above the volume average $\langle L_p \rangle$, and the incident intensity level is 6 dB less than $\langle L_p \rangle$. If we assume that the sound pressure level on the measurement surface is not too different from the rigid wall average (which will be true in the worst cases of poor absorbers), the relationship between δ_{pI} and the specimen absorption coefficient α will be[120]

$$\langle \delta_{pI} \rangle \approx 9 - 10 \lg(\alpha)\,\text{dB} \qquad (10.1)$$

For $\alpha = 0.5$, this is 12 dB, and for $\alpha = 0.1$ it is 19 dB. The phase mismatch bias error of most measurement systems will be of the order of 1 dB in the first case, and unacceptably large in the second. An error of 1 dB corresponds to a fractional error in the estimate of absorption coefficient of 26%, which is unacceptable. The problem of bias error is exacerbated by the fact that the absorption capacities of most common architectural sound-absorbing materials are least efficient at frequencies below 250 Hz, and L_ϕ of most intensity measurement systems is also least at low frequencies.

Normal incidence sound absorption coefficients of materials may be measured in an impedance tube by various forms of manipulation of the signals from two microphones; see, for example, Refs 121 and 122. It is of historical interest that two of the pioneers of sound intensity measurement, namely Clapp and Firestone,[4] presented most impressive comparisons between absorption coefficients measured by the standing wave technique, and by means of measuring the sound energy density and sound intensity in the tube: they were able accurately to measure coefficients as small as 0.07 down to frequencies as low as 125 Hz.

Although the application of sound intensity measurement to the determination of the absorption coefficient of materials does not seem likely to supersede the classical reverberation room method, it has potential as a means of investigating the sensitivity of absorber performance to acoustic field conditions, and for the qualification of absorbers installed in studios, music rooms, etc. In this mode, the Institute of Sound and Vibration Research (ISVR) was able, extremely rapidly, to locate behind the wall covering of a broadcasting studio, a number of resonator absorbers which were not working properly: functioning resonators reveal their presence by the unusually high ratio of mean square particle velocity to pressure in the vicinity of the mouth. In another consulting exercise, ISVR were able to demonstrate in a large concert hall that nearly half the sound power absorbed by the unoccupied seating area was disappearing under the seats into a resonating volume. In principle, the sound power absorbed by any component of an auditorium, including individual persons in an audience (of patient disposition, and/or well paid), may be evaluated in this manner.

10.3   MEASUREMENT OF SPECIFIC ACOUSTIC IMPEDANCE

Specific Acoustic Impedance (SAI) is a mathematically complex quantity defined as the ratio of the complex amplitudes of sound pressure and particle velocity of a single frequency component of a sound field at a point in space: $z = P/U$, where P and U are, respectively, the complex amplitudes of sound pressure and a vector component of particle velocity. Naturally, the definition encompasses Fourier components of fields having arbitrary time dependence. SAI is clearly a vector-like quantity because its value depends upon the direction of the associated particle velocity component. Although the measurement of this quantity is not strictly a valid concern of this volume, it is a function of the same two field variables as sound intensity, and therefore intensity measurement equipment can output signals which may be processed to provide impedance data. Another reason why SAI is of interest in relation to sound intensity is that mean intensity can be written as $I = \frac{1}{2}|P^2|/\text{Re}\{z\} = \frac{1}{2}|U|^2\text{Re}\{z\}$. The quantity which is of most general practical importance is the normal surface SAI, which relates the pressure at the surface of a material body and the component of particle velocity normal to the surface: it is employed as a measure of reflective/absorptive material properties.

Direct (non-FFT) intensity measurement systems can output analogue pressure and particle velocity signals which may be fed directly to the input terminals of an FFT analyser; the transfer function P/U is the desired quantity. SAI may be non-dimensionalised by dividing by the characteristic specific acoustic impedance of the fluid $\rho_0 c$.

It is possible to use FFT analysers to generate the complex spectrum of SAI from the output signals of a p-p probe, but only by an indirect procedure requiring manipulation of spectral quantities. The relationship is

$$z/\rho_0 c = \mathrm{i}(kd/2)[(G_{11} - G_{22} + 2\mathrm{i}\,\text{Im}\{G_{12}\})/(G_{11} + G_{22} - 2\,\text{Re}\{G_{12}\})]$$

$$(10.2a)$$

or the alternative form

$$z/\rho_0 c = \mathrm{i}(kd/2)[(G_{11} + G_{22} + 2\,\text{Re}\{G_{12}\})/(G_{11} - G_{22} - 2\mathrm{i}\,\text{Im}\{G_{12}\})]$$

$$(10.2b)$$

where G_{12} is the cross-spectral density of the outputs of a p-p probe having acoustic separation distance d. The expression of eqn (10.2b) generally yields more reliable results than that of eqn (10.2a) because it is less sensitive to noise on the 'input', which is implicitly the pressure difference, and therefore more subject to noise than the 'output' which is proportional to the pressure sum.[123] If the incident sound field is stationary and random, the random error of a determination may be evaluated in terms of the coherence function γ^2. The remarks in Section 7.3.1 about bias error in the determination of γ^2 apply equally here; estimates of SAI in reflective, multi-mode enclosures require adequately fine frequency resolution and reasonably uniform auto-spectra.

In principle, the great advantage of the direct technique is that a specimen may be insonified *in situ*, and a probe traverse may be made to provide a direct determination of the space-average SAI; by contrast, the value of the SAI at any one point on the surface may only be evaluated from eqns (10.2) if the probe is allowed to dwell at each point for a sufficiently long period for G_{12} to be determined with sufficient confidence. As a by-product of these measurement techniques, the sensitivity of the SAI to variation in the form of the incident field provides an indication of the degree to which a material exhibits 'local reaction'.

An example of the application of the direct technique to the measurement of the real and imaginary parts of the *in situ* impedance of a 'Diagon'-type room absorber when installed correctly along a wall–ceiling intersection, and also when placed hard up against a wall, is presented, together with the diffuse incidence absorption coefficient, in Fig. 10.16. In this case the impedance was derived from $H_1 = G_{up}/G_{uu}$, rather than the more reliable estimator $H_2 = G_{pp}/G_{up}^*$, and it shows the characteristic erroneous decrease in the impedance towards zero at low frequencies caused by noise in the velocity channel. In spite of this deficiency, the effect of the compliance of the contained

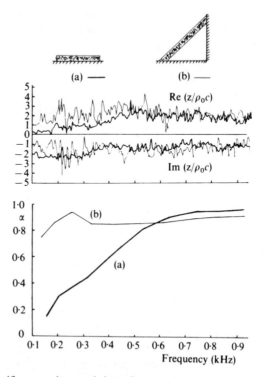

Fig. 10.16. Specific normal acoustic impedance and random incidence sound absorption of two configurations of 50 mm thick fibrous panel.[124]

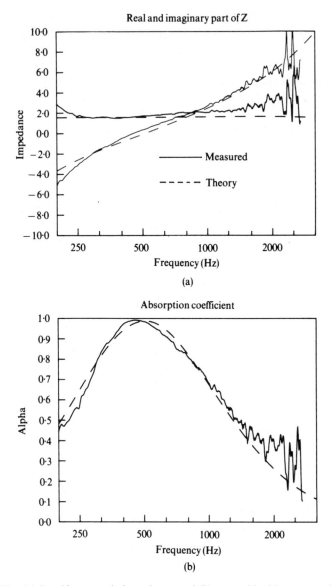

Fig. 10.17. (a) Specific acoustic impedance and (b) normal incidence sound absorption coefficient of a fibrous sheet covered by a perforated sheet—measured by the free field, two-microphone technique (Vigran, T., pers. comm., 1988).

volume, and of the panel vibration modes, is clearly seen in the multi-peaked characteristic of the low frequency reactive impedance curve; the real part is also seen to be increased by the presence of the contained air volume, producing a corresponding increase in low frequency absorption.

Other examples of the application of this technique to free field absorption and impedance measurements are reported in Refs 125–127. An example of application to a Helmholtz resonator is shown in Fig. 10.17 (Vigran, T. E., pers. comm., 1988).

In principle, measurements made on the same material using different forms of incident field should test the degree of 'local reaction' of the absorbent. Deviations from this condition may explain some of the significant discrepancies between 'free field' and impedance tube results seen in the published literature; insufficient attention to the minimisation of bias and random errors may also be partly responsible, together with the necessity to account for the spherical wave spreading from a point source. The freedom from the constraints of an impedance tube, together with the possibility of using any convenient form of sound field, gives valuable potential for use as a quality control method, say at the end of a production line of turbofan inlet acoustic liners, or other commercially fabricated absorber constructions.

10.4 MEASUREMENT OF SOUND POWER TRANSMISSION THROUGH PARTITIONS

Methods of evaluation of the airborne sound energy transmission (insulation) properties of partition structures have application in a wide range of technologies, including those of building, aerospace, railway and road vehicles, surface ships, submarines, industrial machinery and plant, electrical power generation, and turbine and reciprocating engines. Partition insulation performance is usually expressed in terms of the sound power transmission coefficient τ, which is defined as the ratio of transmitted to incident powers. The logarithmic form of this measure is the Sound Reduction Index (R), or Transmission Loss (TL), which are equivalent, and defined as $R(TL) = 10\lg(1/\tau)$ dB. In many practical cases it is the Insertion Loss (IL), which is the difference in received sound pressure level with and without a partition in place, and not the transmission loss, which is significant. The reason why TL is considered to be a more fundamental index of partition performance than IL is that the latter depends upon factors unrelated to the dynamic properties of a partition such as the amount of absorption in a receiving space.

As in the case of absorption measurement, one of the two quantities appearing in the definition of the transmission coefficient, namely the sound power incident upon a partition, cannot normally be determined directly by intensity measurement on the incident face, although attempts have been made to evaluate traffic noise sound power incident upon building facades by covering them in highly absorbent material.[128] This technique is not suitable for use in the source room of a transmission suite because the presence of the absorbent alters the incident sound power. Alternatively, the sound pressure on the surface of a rigid specimen may be measured.[129] If a partition neither dissipates energy nor radiates vibrational energy into connected structures, the net intensity integrated over the incident face equals the transmitted sound power.

The airborne sound transmission properties of a partition are normally measured in the laboratory by placing it between two reverberation rooms; one contains the broad band source(s), and the transmitted sound is measured in a receiving room, calibrated for its acoustic absorption. The incident sound power is inferred from measurements of space-average mean square sound pressure in the source room. This estimate is probably less prone to error than the equivalent estimate of sound power incident upon an highly absorbent specimen in a reverberation chamber. In the standard method of measurement of TL, the transmitted sound power is inferred from the space-average mean square sound pressure in the calibrated receiving room. However, only the total transmitted power is determined, and no information can be derived about the differential transmission properties of the various components of a complex structure, such as a window in a wall: nor can the presence of any particularly transmissive elements, such as small apertures, be detected.

The direct measurement of transmitted intensity offers a number of substantial advantages over the conventional method:[130] (i) the receiving room does not have to be calibrated for its acoustic absorption—nor is such a room actually necessary; (ii) the distribution of transmitted intensity over the surface of the partition may be determined, thereby revealing the presence of weak areas, or leaks; (iii) the sound power radiated by the dividing partition, and by other associated structures, may be separately determined, thereby allowing detection and precise quantification of flanking path transmission. Intensity measurement is particularly effective at exposing the presence of airpath leaks in partitions, because not only does the normal intensity peak in a narrow region, but the tangential intensity abruptly changes sign as the probe passes across a leak. It is, however, quite difficult to evaluate the transmitted power because the spatial gradients of normal intensity are very high. The application of sound intensity measurement to the determination of transmitted power is demonstrated by Cops and Minten[131] and van Zyl and Erasmus[132] from whom Fig. 10.18 was taken. An example of the distribution of sound intensity over a window at 2 kHz is shown in Fig. 10.19.[133]

The transmitted sound intensity distribution is normally measured on a surface parallel to a partition. Measurements on the peripheral faces of the enclosing surface must not be neglected; a significant proportion of the transmitted power may be transported through these faces, especially at frequencies in the neighbourhood of the critical frequency of a panel structure. Some systematic investigations have been made of the influence of measuring distance and sample point density on the accuracy of estimated transmission loss, for example by Guy and De Mey.[134] There can be no universally applicable recommendation for these parameters because the spatial variation of intensity in the near field of vibrating structures depends strongly on the degree of non-uniformity of the structure and on the ratio of measurement

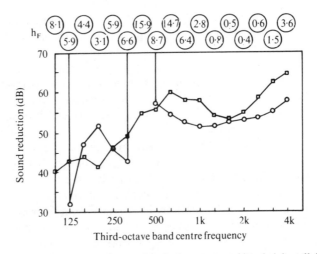

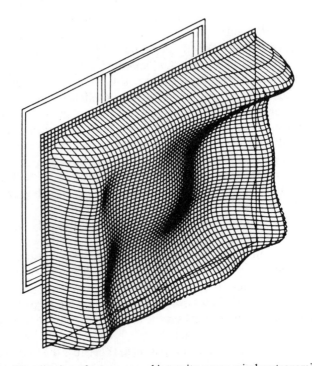

Fig. 10.18.   Sound reduction index R of a window mounted in a brick wall; h_F = flanking power/direct power: ○ conventional method; □ intensity-based method.[132]

Fig. 10.19.   Distribution of mean normal intensity over a window transmitting sound in the 2 kHz 1/3 octave band.[133]

frequency to critical frequency of the partition material. In general, the density of measurement points must increase in proportion to the critical frequency of the partition, particularly if the partition incorporates stiffening frames. The measurement distance must be chosen as a compromise between attempts to avoid the recirculation regions in the very near field and a desire to minimise the adverse influence of the reverberant field in the receiving room: distances of between 50 and 200 mm are commonly employed.

Comparisons of results obtained by the point sampling and scanning methods have generally shown close agreement; however, it is not wise to measure very close to a partition surface, especially at discrete points, because of the very complex form of near field intensity patterns. The presence of a human operator may be beneficial in providing screening and absorption.

The relationship between TL and the measured sound pressure and transmitted power levels is

$$TL = \langle L_p \rangle - L_w + 10 \lg S - 6 \, \text{dB} \qquad (10.3)$$

where S is the area of the partition and $\langle L_p \rangle$ and L_w are respectively the space-average sound pressure level in the source room and the transmitted sound power level.

An approximate expression for the average pressure-intensity index in the transmitted field close to the surface of a partition is

$$\langle \delta_{pI} \rangle \approx 9 + 10 \lg(S/A) \, \text{dB} \qquad (10.4)$$

where A is the absorption of the receiving room. If we assume δ_{pI0} for a typical measurement system to be 18 dB, the maximum value of S/A for a p-p system bias error of less than ± 1 dB due to phase mismatch is 1·6. This ratio is clearly greatest in those test configurations where the partition forms the complete dividing wall. The total receiving room surface area will then be of the order of $6S$, and $A \approx 6S\bar{\alpha}$, where $\bar{\alpha}$ is the average wall absorption coefficient. Thus, $S/A \approx 1/6\bar{\alpha}$, and for $S/A < 1\cdot6$, $\bar{\alpha}$ must be greater than 0·1. This is usually the case, except at low frequencies (typically $< 200 \, \text{Hz}$). Even values of S/A as low as 1·4 have been shown to produce significant bias errors.[135] In any case, the introduction of a few square metres of absorbent blanket into a receiving room will usually ensure that the bias errors are negligible, at least for measurements made on uniform partitions. Where a construction comprises components of widely differing TL, the value of δ_{pI} on the 'better' components will exceed that given by eqn (10.4). Experience suggests that it is desirable to limit the measurement surface-average value of δ_{pI} to 10 dB or less.

The greatest problem in the *in situ* determination of the sound transmission properties of walls, floors and ceilings in buildings, or other multi-path structures such as ships, is that of separating the contributions to the received sound of the many possible flanking paths of transmission from the source region. In principle, intensity measurement offers the facility for the separate evaluation of the contributions from each of the surfaces bounding a receiving space, by enveloping each in an intensity measurement surface. Since the intensity generated by flanking surfaces is generally much smaller than that of the primary partition, the average pressure-intensity index is much larger, and the problem of bias and random errors in the evaluation of flanking transmission is generally much more severe than with the primary partition measurement, assuming that the latter is the major transmitting element. The situation is reversed in cases where a flanking path dominates. Errors may in some cases be so large that the rank ordering of surfaces may be faulty. A successful application of intensity measurement to the detection and evaluation of flanking paths in a building is described in Ref. 136.

Detailed analysis of the acoustic characteristics of transmission suites indicates an additional influence of the receiving room field on apparent transmission loss when the receiving room side of a partition is itself significantly absorbent.[137] The net measured normal intensity will equal the radiated (transmitted) component minus that absorbed by the surface from the reverberant field. The latter will increase with receiving room sound pressure level, and with receiving room reverberation time. Of course, this mechanism will also operate in practical installations, but not usually to the same degree because most rooms in buildings are more absorbent than laboratory receiving rooms. In order to minimise errors associated with this mechanism, it is advisable to introduce auxiliary absorption into a receiving room, so that a realistic balance between radiated and absorbed power is achieved. This requires that the ratio S/A is the same in laboratory and field situations, where A includes the absorption of the test partition. In Ref. 137, and in another comparative study of conventional and intensity-based TL measurement techniques,[135] the authors recommend the application of the 'Waterhouse correction' for the boundary effects on energy density distribution in a room (see also ISO 3741):

$$TL = \langle L_p \rangle - L_w + 10 \lg S - 6 + 10 \lg(1 + \lambda S_1/8V_1) \, dB \qquad (10.5)$$

where λ is the acoustic wavelength at the band mid-frequency, and S_1 and V_1 are respectively the source room surface area and volume.

A recent collaborative programme of measurements carried out by three laboratories with the purpose of establishing the precision of the scanned intensity method in comparison with the traditional pressure-based method has shown that agreement is generally excellent.[138] A similar conclusion was drawn from a parallel multi-team investigation.[139] The spread of transmission loss data obtained by the different teams during a field test was, however, considerable (up to 10 dB at 125 Hz); it was tentatively attributed to the differing forms and deployments of the test sources. The differences between the intensity-based estimates were generally less than those based on the standard method of ISO 140:4. Application of the Waterhouse correction, based on the receiving room dimensions, to the intensity-based estimates of transmission loss produced a systematic overestimation of the weighted sound reduction index R_w (ISO 140:3) with a standard deviation of 1 dB, whereas its omission produced a small systematic underestimation. It was concluded that the intensity-based method is more reliable for the determination of the sound insulation produced by high performance partitions such as double windows because of its immunity to flanking transmission.

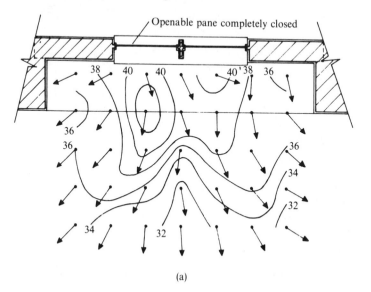

(a)

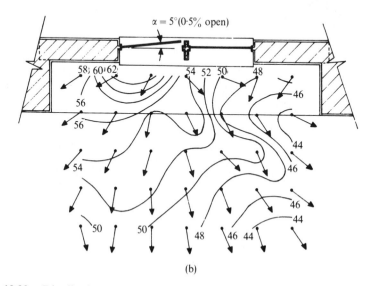

(b)

Fig. 10.20. Distribution of mean intensity and intensity levels in the median plane of a window transmitting noise in the 1 kHz 1/3 octave band: (a) closed window; (b) window opened 5°.[128]

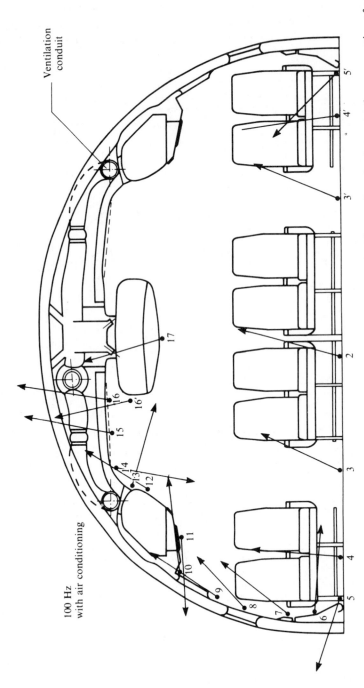

Fig. 10.21. Projection of sound intensity vectors in the 100 Hz 1/3 octave band measured at the numbered points onto the cross-section of an aircraft with the air-conditioning system operating: vector scale proportional to sound intensity level.[142]

Ventilation conduit

100 Hz with air conditioning

Brüel & Kjær

Averaging: 16 sec. lin Ref. Level: 60 dB Rec. No.: _____ Sign.: J.H./B.G. Date: 10/5/82 Weight. Netw.: ☐☐☐☐☐☐☐☐

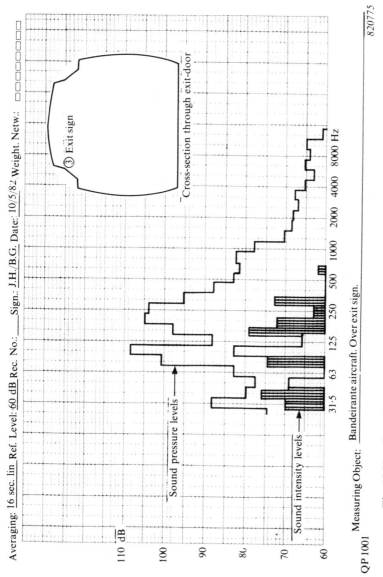

QP 1001 Measuring Object: Bandeirante aircraft. Over exit sign.

820775

Fig. 10.22. Sound pressure level and sound intensity level measured over exit sign.[142]

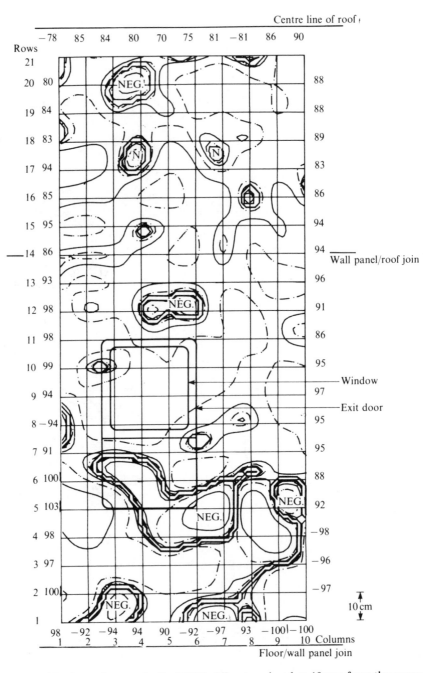

Fig. 10.23. Intensity map in the 100 Hz 1/3 octave band at 10 mm from the source surface; iso-intensity contours at intervals of 6 dB.[142]

Among the many applications of sound intensity measurement to the determination of building partition sound transmission properties are a number which address specific aspects of design, operation and installation. The effects of window opening were investigated by Migneron and Asselineau,[128] and examples of the intensity field are shown in Fig. 10.20. Guy and De Mey[134] investigated the effect of absorbent reveals (aperture surfaces) on the transmission loss of single glazing. They observed significant increases in *TL*, particularly above 400 Hz, and concluded that the mechanism was not the reduction of the sound power radiated by the partition but the subsequent absorption by the reveal.

Cops and Minten[131] and Halliwell and Warnock[135] among others, have used the intensity method to investigate the influence on *TL* of the placement of a partition within the thickness of an aperture between two reverberation rooms—the so-called 'niche effect'.

The intensity-based technique has been successfully applied to the field evaluation of the *in situ* transmission loss of saddle roof constructions.[140]

Intensity measurement has been applied to the determination of the airborne sound transmission properties of many non-building structures, such as car bodies, machinery enclosures and aircraft fuselages, e.g. Refs 141 and 142. In these cases, the primary interest is in the distribution of the transmitted power for the purpose of specifying noise control measures. Reference 142 presents results of intensity surveys made in a propeller-driven aircraft. The intensity distributions shown in Figs 10.21–10.23 are typical of those found in the confined reflective spaces of vehicle compartments: intensity vector distributions often fail to reveal distinct energy flow patterns; the surface field exhibits a mixture of sources and sinks; and the pressure-intensity index is often very high, exceeding 20 dB at some frequencies in the cases cited above. Interpretation of such distributions for the purposes of selecting noise control measures is fraught with uncertainty. However, high frequency transmission paths such as leaky seals can be readily detected.

10.5 RADIATION EFFICIENCY OF VIBRATING SURFACES

Radiation Efficiency, also known as Radiation Ratio, is a measure of the effectiveness with which a vibrating surface generates sound power. It is defined by

$$\sigma = W/\rho_0 c S \langle \overline{v_n^2} \rangle \tag{10.6}$$

where W is sound power radiated by a surface of area S, and $\langle \overline{v_n^2} \rangle$ is the space-average mean square normal velocity of the surface. This index is really only significant when the distribution of mean square velocity is reasonably uniform. If the surface is that of a plate or shell of uniform thickness, vibrating in flexure, σ is proportional to the ratio of radiated sound power to mechanical vibration energy, thus:

$$\sigma = (W/E)(h/c)(\rho_m/\rho_0) \tag{10.7}$$

in which E is the vibrational energy, h is the thickness and ρ_m is the material density.

The sound power radiated by a structure may be determined by intensity measurement, and the surface velocity may be measured with accelerometers, optical or ultrasonic transducers, or an intensity probe. The latter measurement is necessarily approximate because the particle velocity can only be measured close to, but not on, a surface. A p-u probe indicates v_n directly, as does the analogue velocity output signal from a direct p-p system. The spectral outputs of an FFT p-p system may be processed to give the approximate auto-spectrum of surface velocity:

$$G_{vv} \approx (1/\rho_0\omega d)^2 [G_{p1p1} + G_{p2p2} - 2\,\mathrm{Re}\{G_{p1p2}\}] \tag{10.8a}$$

This estimate of the velocity auto-spectrum in the near field of vibrating surfaces is subject to considerable bias error in the presence of the evanescent near fields associated with such surfaces, as suggested by eqn (5.35b). It is also subject to random error in the presence of extraneous acoustic fields.[143] Alternative spectral formulations have been devised by Steyer in an attempt to reduce these errors:[144] two examples are

$$G_{vv} \approx (1/\rho_0\omega d)^2 \{[(G_{p1p1} - G_{p2p2})^2$$
$$+ 4(\mathrm{Im}\{G_{p1p2}\}^2)]/[G_{p1p1} + G_{p2p2} + 2\,\mathrm{Re}\{G_{p1p2}\}]\} \tag{10.8b}$$

and

$$G_{vv} \approx (1/\rho_0\omega d)^2 \{[(G_{p1p1} + G_{p2p2} - 2\,\mathrm{Re}\{G_{p1p2}\}]$$
$$- 4|G_{p1p2}|^2(1/\gamma^2 - 1)/[G_{p1p1} + G_{p2p2} + 2\,\mathrm{Re}\{G_{p1p2}\}]\} \tag{10.8c}$$

The expression in eqn (10.8b) contains the sum of two terms, the first being related to the reactive intensity component and the second to the active component: it is purported to be less sensitive to error than eqn (10.8a) when uncorrelated random noise is present in the two pressure signals. Equation (10.8c) is proposed only for 'very near field' velocity estimations.

The advantages and disadvantages of these various formulations have been demonstrated in a comprehensive series of experiments by Steyer.[144] In general, the estimate provided by eqn (10.8a) was found to be less sensitive to interference by extraneous acoustic fields, but the other two gave better dynamic range. All the estimates were found to be subject to considerable error when used to evaluate the vibration velocity of a stiffened vibrating plate, especially at frequencies close to the critical frequency of the plate.

The radiation efficiency is given by

$$\sigma = (kd)\langle \mathrm{Im}\{G_{21}\}\rangle/\langle G_{11} + G_{22} - 2\,\mathrm{Re}\{G_{12}\}\rangle \tag{10.9}$$

in which $\langle\ \rangle$ indicates space-average over S. This estimate is subject to bias and random errors in both the numerator and denominator. The uncertainty in an estimate of the space-average mean square velocity of the radiating surface depends strongly on the nature of the vibration field and on the signal processing procedure employed.

A number of examples of this application have been published,[123,143-145] but the results generally indicate that very great care must be taken to minimise the errors due to probe orientation, spatial sampling technique and extraneous noise sources. A comparison between velocity spectra measured at 90 points on a 4 mm thick steel plate structure by accelerometer and p-p intensity probe is shown in Fig. 10.24.[143] Figure 10.25 shows the effect of an extraneous source on the estimate of radiation efficiency. Loyau[146] demonstrates the superiority of a three-microphone estimate of surface velocity over the conventional two-microphone technique.

It should not be forgotten that sound radiated from regions of an extended source which do not form part of the measured surface (say a machine cover panel), contributes to the measured surface pressure, and also to the measured normal velocity. Pascal also shows that the extraneous velocity component

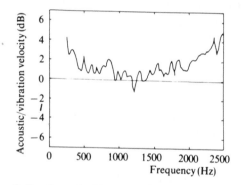

Fig. 10.24. Ratio of vibration velocities normal to a vibrating plate as measured with a p-p probe and an accelerometer in the presence of extraneous noise.[143]

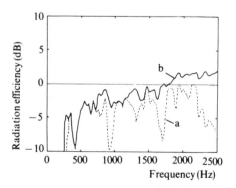

Fig. 10.25. Estimates of radiation efficiency using the p-p system: (a) in the presence of extraneous noise; (b) in the absence of extraneous noise.[143]

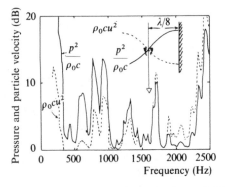

Fig. 10.26. Effect of extraneous noise on estimates of near field sound pressure and particle velocity.

may be suppressed if the measurement is made at a distance closer than $\frac{1}{8}$th wavelength, which is, in fact, a typical microphone separation distance. The reason for this restriction is the presence at the surface of an interference pattern of extraneous noise (Fig. 10.26). This observation supports that of Hübner[84] that the contribution to the measured normal intensity component of extraneous sources is small in the vicinity of a solid surface.

10.6   TRANSIENT NOISE SOURCES

Sound sources such as mechanical impacts, electric sparks, gun shots, etc., produce transient (non-stationary) sound fields. In reverberant surroundings, sound energy will continue to flow long after the source mechanism has ceased to operate. Direct methods of sound intensity measurement are entirely suitable for evaluating instantaneous sound power flux density, and appropriate time integration may be applied to yield the sound energy flux density through a surface during any selected period of time. When measurements of the temporal evolution of a transient intensity field are made with a p-p probe, the complete expression in eqn (5.5) must be evaluated, rather than the condensed form of eqn (5.6).

According to eqn (4.11b), the rate of increase of energy density of a region of fluid is equal to the rate at which work is done on unit volume of fluid by active source mechanisms minus the net rate at which sound energy leaves the region through its boundaries. When the sources are stationary, the time-average rate of change of energy density is zero, and the source sound power is equal to the surface integral of the normal component of intensity over an enveloping surface, as expressed by eqn (9.2). In cases of non-stationary sources, eqn (4.11b) may be integrated with respect to time to express the change in energy density over any period of time in terms of the sound energy injected per unit volume by active sources and the time integral of the divergence of the instantaneous intensity vector. Application of Gauss's integral theorem shows that the difference between the increase of the sound energy of an arbitrary volume of fluid during an arbitrary period of time is equal to the difference between the sound energy injected into it by active sources and the sound energy which leaves the volume through its boundaries during that period (assuming an absence of dissipative agents in the volume). Thus the time integral of the normal intensity over an enveloping surface does not necessarily equal the sound energy generated by the enveloped sources during the period of integration. If, however, the period of integration extends from any instant prior to the activation of a transient source until the time when the sound field within the bounding surface has *ceased to exist* (due to radiation to 'infinity', or the action of dissipative mechanisms in the region external to the enveloping surface), the net increase of fluid sound energy is zero. Hence the time integral of the surface integral of normal surface intensity equals the total sound energy generated by the sources operating within that surface.

The fact that sound fields generated by transient sources in a reverberant space may last for several seconds presents no problems for direct intensity measurement systems. If FFT-based intensity analyses are applied to transient measurements, it has been shown in Section 5.4 that a cross-spectral energy expression equivalent to eqn (5.19) may be evaluated, provided that each pressure-time history is Fourier transformed in its entirety; no time-windowing or time-averaging may be applied. Dedicated FFT analysers are generally not capable of performing a single transform over periods of seconds, and therefore special-purpose computer processing is necessary.

Implementation of the transient form of the surface integral theorem presents a number of additional practical difficulties. The sound energy density may only be evaluated at one point on the measurement surface at a time. Consequently, evaluation of the integral requires repeated activation of the source, which should ideally produce identical events. If repeatability is not precise, some form of average must be taken at each measurement point. The individual events must be separated by sufficient time for the associated sound field to decay into the measurement noise floor. The presence of extraneous noise generated by other sources can produce serious errors, particularly if it is itself non-stationary. The large dynamic range of transient signals puts strains on signal conditioning and processing systems.

Very few examples of the application of sound intensity measurement to the determination of radiated sound energy and source mechanism investigation of transient machinery sources have been reported. There are a number of practical reasons for this paucity, as detailed by Watkinson:[147]

(i) It is necessary to make a number of measurements at each measurement point on an enveloping surface, in order to obtain an accurate estimate of the mean of repeated, but not identical, events.

(ii) The presence of background noise can seriously affect the estimate of time-integrated intensity. If this noise is also non-stationary, accurate estimates of radiated energy are impossible.

(iii) Random errors due to instrumentation deficiencies (e.g. tape recorder transport irregularities) do not average out as they tend to do with continuous signals.

(iv) The large dynamic range of transient signals places greater demands on instrumentation than continuous signals. For example, at least 12 bit a.d.c. quantisation is necessary.

Figures 10.27(a–c) show the results of measurements on a drop hammer),[49] and Figs 10.28(a, b)[65] show the distribution of 1/3 octave band energy flux close to one side of a model punch press. The presence of regions of negative flux indicates that even in transient sound fields sound energy radiated by one part of a source can flow back in elsewhere. The complicated nature of transient source intensity fields is illustrated by Fig. 10.29,[147] in which the evolution in time of the intensity field of a transiently accelerated baffled piston is displayed. Measurements by Alfredson[148] clearly reveal the relationship between sound energy radiation and various mechanical events in the cycle of a punch press.

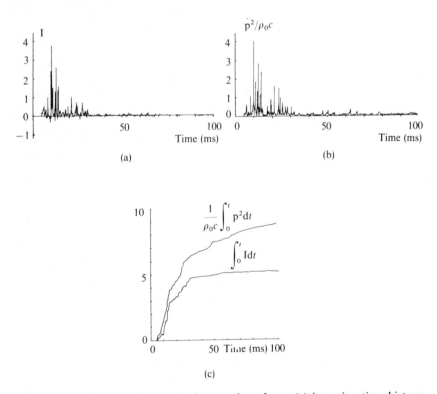

(a)

(b)

(c)

Fig. 10.27. Sound intensity at 1 m from a drop forge: (a) intensity–time history; (b) time history of 'pseudo-intensity' $p^2/\rho_0 c$; (c) time integrals of intensity and 'pseudo-intensity'.[49]

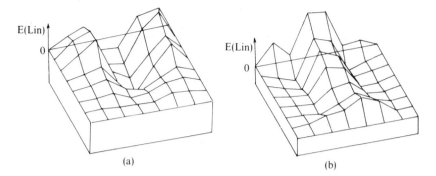

Fig. 10.28. Sound energy flux distributions from the side of a model punch press in 1/3 octave bands: (a) 163 Hz; (b) 200 Hz.[65]

10.7   ACTIVE NOISE CONTROL SYSTEMS

A number of different principles are employed to control noise by active means; one of these is the active optimisation of sound absorption elements.[149] The effect of an optimised sound absorber on incident plane waves is shown in Fig. 10.30. In this case particularly, but also in the general field of active noise control, it is of interest to investigate the sound powers radiated (or absorbed) by primary and secondary sources, and how they are influenced by source interaction. Two examples of the theoretical sound intensity vector distributions in a two-dimensional active control system are shown in Figs 10.31(a, b).[150]

10.8   APPLICATIONS TO HOLOGRAPHY

10.8.1   Broad band acoustical holography by intensity measurement (BAHIM)

Near field Acoustical Holography (NAH) is most commonly used to identify principal sources of far field radiation by means of field measurements made close to the source surface.[151] It is based upon the property of a three-dimensional sound field that the whole field can be derived from a knowledge of the value of the sound pressure or its normal derivative on any infinitely extended surface which is a constant parameter surface in any of the co-ordinate systems in which the scalar wave equation is separable (e.g. rectangular, cylindrical): the Rayleigh integral is an example of this principle. By means of this principle, measurements of pressure or normal particle velocity on a suitable surface in the vicinity of a source of sound may be used to evaluate either the normal surface velocity of the source surface or the far field radiated by the source. Naturally, measurements can be made only over the extent of an aperture of finite dimensions and therefore assumptions have to be made

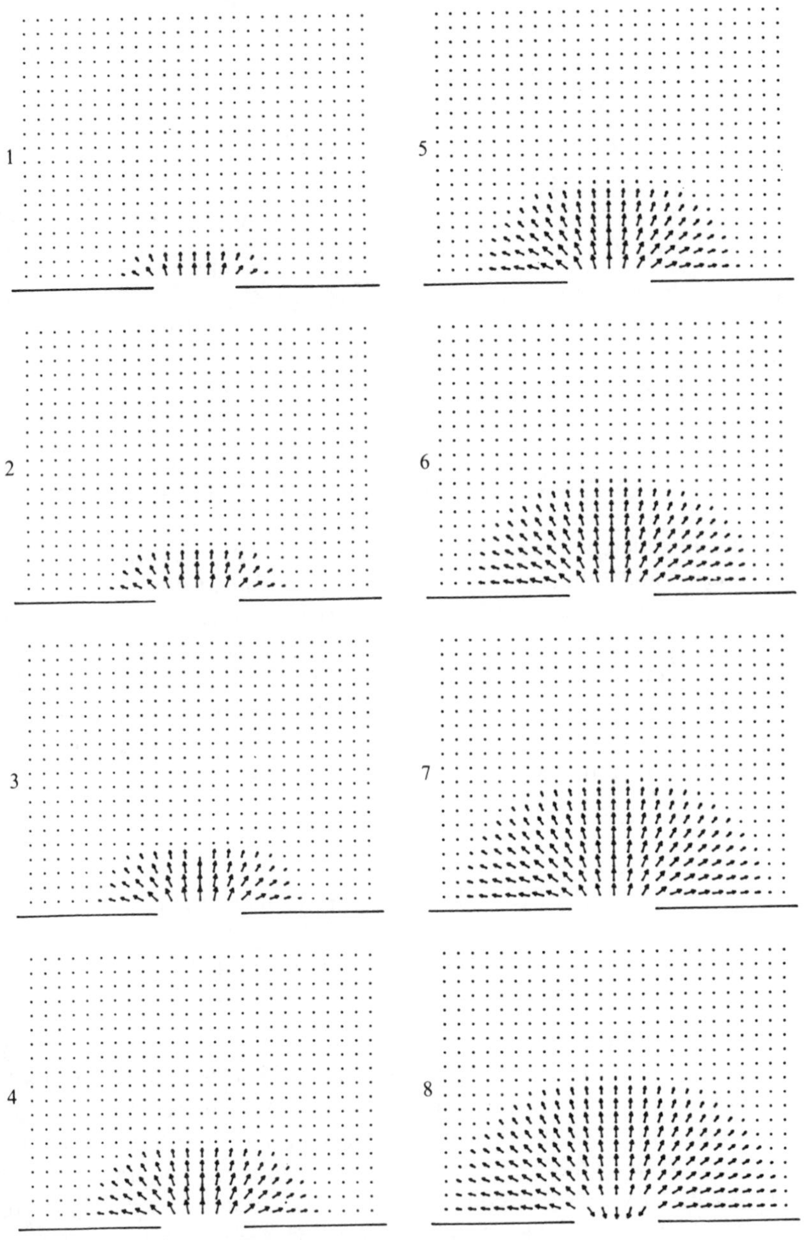

Fig. 10.29. Instantaneous intensity vectors in the field of a circular piston which executes a single cycle of sinusoidal motion in 0·5 ms; intervals between figures = 0·025 ms.[65]

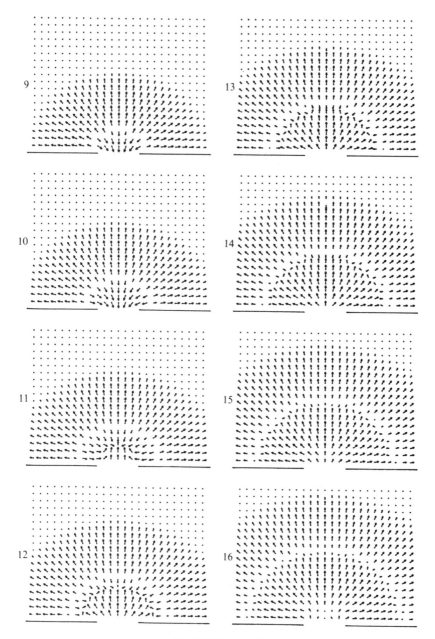

Fig. 10.29—*contd.*

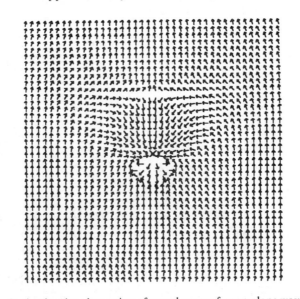

Fig. 10.30. Optimal active absorption of sound energy from a plane wave by a pair of point monopoles.[149]

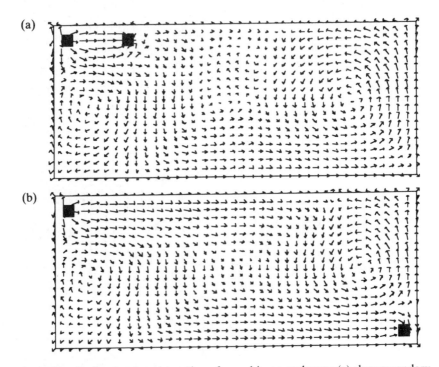

Fig. 10.31. Optimal active absorption of sound in an enclosure: (a) close secondary source; (b) remote secondary source.[150]

about the field on the extension of the surface beyond the aperture. Single surface holograms cannot distinguish between sound waves arriving from different sides of the surface, and therefore the single surface technique is restricted to conditions in which the influences of reflections of radiated sound back towards the source, and of extraneous sound generated by other sources, are negligible.

BAHIM is a technique by which the two orthogonal components of mean intensity *tangential* to a hologram (measurement) plane are sampled at points on a square grid and converted to distributions of pressure amplitude and phase, from which the whole sound field can be reconstructed, subject to the limitation imposed by wavenumber aliasing (Shannon criterion for spatial frequency analysis).[152] The relationship between mean intensity and complex pressure $P \exp(\phi_p)$ in a harmonic field, for which BAHIM is formulated, is given by eqn (4.17), namely

$$I(x) = -(1/2\omega\rho_0)[P^2(d\phi_p/dx)] \tag{10.10}$$

In a time-stationary, aperiodic field, the average phase gradient in the resolution band is given by eqn (10.10) in terms of the spectral densities of pressure and intensity. The spatial distribution of pressure phase is obtained by numerical spatial integration of eqn (10.10) in the two orthogonal directions which removes the indeterminacy associated with a single direction. Phase is thus obtained without a need for a reference signal, subject, of course, to steadiness of source operation. Once the phase distribution is obtained, the standard NAH procedure may be followed to reconstruct the whole pressure field from which any quantity of interest, such as particle velocity, energy density and reactive intensity, may be derived. The pressure phase distribution on the hologram plane depends upon the coherence between the sources generating that distribution; consequently, projection of the measured field onto a source surface will only yield physically significant results if the field is generated predominantly by a coherent source.

An application of BAHIM to the identification of the principal regions of sound power radiation of an air compressor[152] demonstrates that the ability to estimate the distributions of active and reactive intensity on a plane surface very close to the source surface, without the problems of access and measurement error associated with direct measurements in the very near field, can be more effective as a means of locating regions of strong power radiation than the conventional measurement of intensity on a more remote surface. Figures 10.32 and 10.33 present the distributions of active and reactive intensity on the intensity measurement plane at 120 mm from the source surface and on the projection plane very close to the source surface. Because of interference effects in the geometric near field, the former give misleading indications of the principal source regions; those provided by the latter are far more credible. It should be noted that the sound field generated by the compressor was dominated by tonal components which are naturally associated with coherent sources.

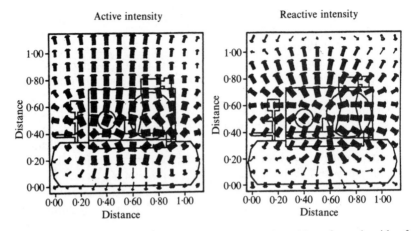

Fig. 10.32. Conventional intensity measurements made at 12 cm from the side of a motor at 300 Hz.[152]

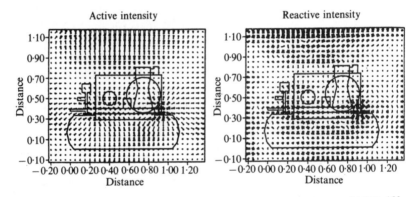

Fig. 10.33. BAHIM results at 12 cm from the side of the motor at 300 Hz.[152]

It is also worth noting that it is not necessary to evaluate the mean intensity, which is subject to unavoidable and unknown finite difference error, in order to estimate the local phase gradient, since the relative phase of the pressures measured by the two microphones is given by

$$\sin(\Delta\phi) = \mathrm{Im}\{G_{12}\}/|G_{12}| \tag{10.11}$$

from which the phase gradient $\Delta\phi/d$ can be evaluated. If the pressure spectrum is not dominated by tonal peaks but is rather smooth, and the phase difference varies very rapidly with frequency, the estimate of $\Delta\phi$ depends strongly on the analysis frequency resolution; in this case it is likely that the sound field is generated by a number of uncorrelated sources, and BAHIM may not be effective.

10.8.2 Spherical acoustical holography using sound intensity

The technique of spherical acoustical holography using sound intensity is based upon eqn (10.10) but, in contrast to BAHIM, only the component of mean intensity directed along parallels is determined[153]. It is not necessary to determine the meridional component to eliminate the indeterminacy of phase derived from integration of its gradient because all meridians meet at common points on the poles of a sphere. It is shown in Ref. 153 that phase mismatch greater than field phase difference does not affect the accuracy of the field projection.

10.9 INTENSITY AND RECIPROCITY

Noise control engineers frequently need to identify those regions of a vibrating source of sound which most strongly contribute to the sound pressure at one or more *specific points*; for example, at the ear of a machine operator in a factory or at the ears of passengers in a vehicle. It has been emphasised in Section 10.1.2 that there is often little direct relationship between sound power and sound pressure in sound fields generated in reflective enclosures. It is also impossible to attribute intensity at any field point remote from a source surface to any particular source or source region in cases where many different agencies contribute significantly to the sound field. Consequently, sound intensity measurement *per se* is not likely to be of much diagnostic assistance in such cases.

The relationship between the harmonic surface vibration of a small area of a surface and the sound pressure at a specific point in the sound field generated by that vibration may be expressed in terms of the Green's function;[154] this is a transfer function which expresses the magnitude and phase of the sound pressure produced by unit volumetric acceleration of the source surface. Green's functions for any vibrating surface operating in any acoustical environment may be determined by applying the principle of acoustic reciprocity. A calibrated omni-directional sound source is placed at the desired observation point and transfer functions are measured between the source strength and the sound pressures at a grid of points distributed over the inert surface of the source: these are the desired Green's functions.[155] The normal acceleration distribution of the operating system is appropriately sampled and the values are combined with the Green's functions to yield an estimate of the mean square sound pressures at the points of concern.

The most common forms of vibration transducers, namely accelerometers, sense vibration at a 'point'. This is not ideal for reciprocity applications in which volume acceleration is the desired quantity; spatial aliasing can produce false estimates.[156] Although special-purpose capacitative volume acceleration transducers have been devised,[157] they are not suitable for general purpose investigations on vehicles, machinery and industrial plant, of which the surfaces may be physically inaccessible, hot or dirty or may not readily accept the attachment of an accelerometer; for example, carpets and lightweight vehicle trim.

In order to overcome such problems, Verheij has introduced a method wherein it is assumed that the vibrating surface may be represented acoustically as an array of *uncorrelated* monopoles associated with individual subdivisions (segments) of the total surface.[158] On the basis of this assumption it is necessary only to evaluate the magnitude of each Green's function and not the phase. The relationship between the sound power radiated by a monopole and its volumetric strength is well known, and hence estimates of the vibrational source strengths of each 'monopole' are derived from estimates of sound power radiated by each segment as determined by means of scanned near field intensity measurements; corrections are made to account for the fact that the 'monopoles' are not radiating into free space. The resulting distribution of 'monopole' strengths is combined with the magnitudes of the corresponding Green's functions to produce an estimate of the total mean square sound pressure at an observation point together with the contributions of each surface segment.

This method has been shown to be easy to apply and very effective,[158,159] although the assumption of uncorrelated monopoles does not apply in cases where the vibration is dominated by single resonant modes or is tonal and highly coherent: in such cases the relative phases of the various segments influence the radiated power (see Section 9.2). However, there are many practical examples of broad band vibrational sources of sound which exhibit only rather small spatial correlation areas, even if the vibration is fully coherent at each frequency. A good example is provided by the internal combustion engine of which the deterministic vibration is dominated by multiple harmonic components: in a frequency band containing a number of these components the relative phases of vibration of each harmonic at different points on the engine surface take values which are widely distributed over 2π radians; hence the associated frequency band correlation is poor even though the vibration is deterministic.

10.10   EVALUATION OF SOUND FIELDS IN ROOMS

The measurement of the instantaneous sound intensity vector and its temporal envelope forms a central component of a comprehensive room acoustic indicator measurement and evaluation system developed by Guy and Abdou.[160] Figure 10.34 shows the temporal and directional structure of the sound intensity vector in the 500 Hz 1/3 octave band in a transient sound field in a reverberant laboratory. The resulting data may be used to evaluate various indicators of sound field diffusion and have the potential to provide a most valuable link between the physical behaviour of a sound field and its subjective quality.

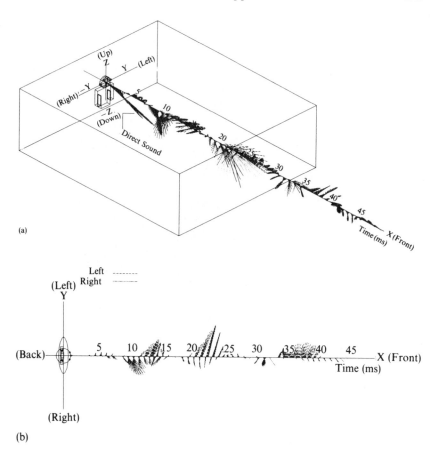

(a)

(b)

Fig. 10.34. Directional information from a measurement in a reverberant field at 500 Hz vs time: (a) isometric illustration (intensity W m^{-2}); (b) directional and temporal structure of left and right reflections in the horizontal plane.[160]

10.11 MISCELLANEOUS APPLICATIONS

Sound intensity measurement has been applied to a number of studies outside the general field of noise measurement and control. The mechanisms and characteristics of sound generation and radiation by musical instruments have revealed some interesting behaviour. Figure 10.35(a–d) shows intensity fields of a violoncello,[161] Figs 10.36(a) and (b) show radiation from a double bass,[162] and Figs 10.37(a–c) show the radiation pattern of a wind instrument.[161] The intensity field of a guitar is illustrated in Ref. 163.

The beneficial effect of reflectors in increasing the flow of sound energy from a stage into an auditorium has been demonstrated using a four-microphone intensity probe.[164]

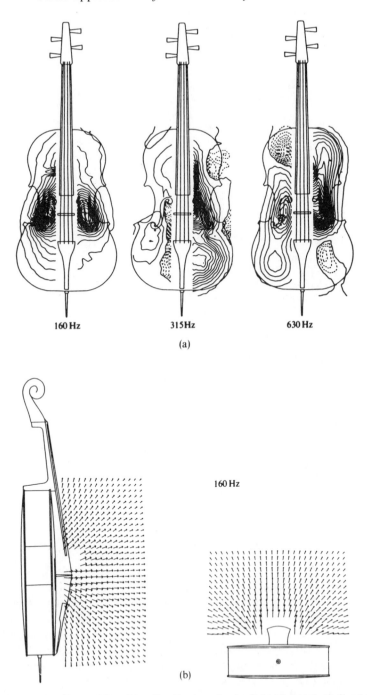

Fig. 10.35. Mean sound intensity distributions in the field of a violoncello: (a) iso-normal intensity contours (broken line negative); (b–d) mean intensity vectors in the median plane and in the plane of the bridge.[161]

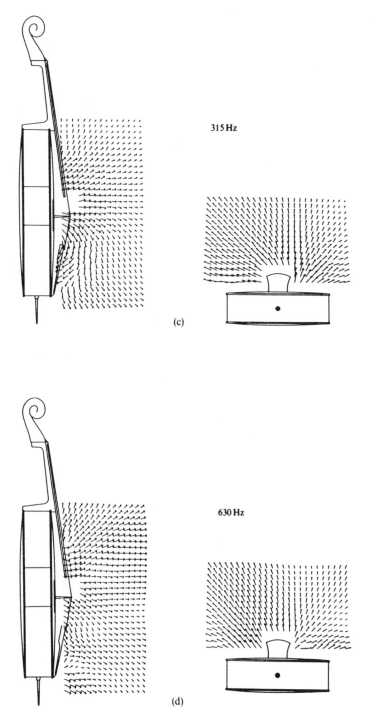

315 Hz

(c)

630 Hz

(d)

Fig. 10.35—*contd.*

Fig. 10.36. Mean sound intensity distributions in the bridge plane of a double bass: (a) 98 Hz; (b) 230 Hz.[162]

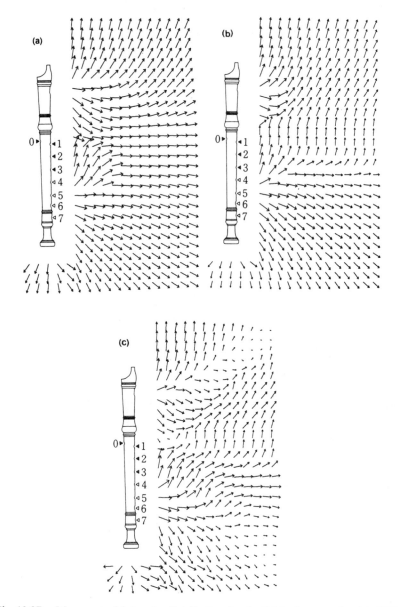

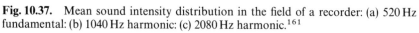

Fig. 10.37. Mean sound intensity distribution in the field of a recorder: (a) 520 Hz fundamental: (b) 1040 Hz harmonic: (c) 2080 Hz harmonic.[161]

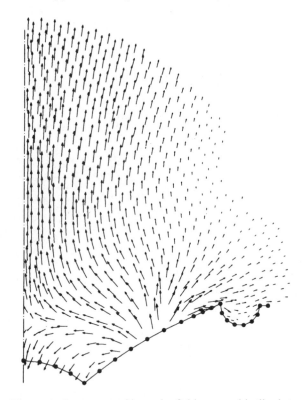

Fig. 10.38. Theoretical mean sound intensity field near an idealised structural model of a loudspeaker cone at 7200 Hz.[165]

A theoretical investigation of the intensity field near a loudspeaker[165] suggests that acoustic power may be absorbed by certain regions of the vibrating surface, as indicated by Fig. 10.38, which also reveals side lobe development of the radiated field.

CHAPTER ELEVEN

Sound intensity in flow ducts

11.1 INTRODUCTION

The process of propagation of sound within ducts is of practical concern in many areas of engineering acoustics and noise control, among which may be mentioned heating, ventilating and air conditioning (HVAC) systems, IC engine exhausts, gas turbine and turbo-jet intakes and exhausts, industrial pipework, wind tunnels, and fluid distribution networks. In ducts of small cross-sectional area, such as car exhaust pipes, sound propagates in the form of plane waves over most of the frequency range of practical interest, and theoretical and experimental techniques for the analysis of the behaviour of such systems have been under development for many years.[166] However, in larger systems, higher order (transverse) modes of propagation can also transport energy. Estimates of sound energy flux cannot then be made unless the duct is terminated artificially with an anechoic device, and then only by the employment of elaborate arrays of microphones, and complex signal processing techniques. Measurement of the axial component of sound intensity in such fields could, in principle, overcome this fundamental difficulty. We shall see, however, that formidable technical difficulties are created by the presence of mean fluid transport in a duct—difficulties which are currently the subject of basic research.

Of the many areas of engineering interest mentioned above, the only one in which the development of practical sound energy measurement techniques for multi-mode fields is likely to be feasible in the near future is that of HVAC systems; mean air speeds are relatively low ($< 30\,\mathrm{m\,s^{-1}}$), air temperatures and static pressures are close to atmospheric, ducts are large and access is reasonably simple, and there is an economic incentive to make noise control systems more efficient, and therefore less costly. At present, the procedures used to calculate the distribution of sound power in a multi-branched duct-work system, and hence to specify the performance required of the noise control components, are based upon a mixture of simplistic theory, empirical data and experience. In the hands of an experienced designer, they usually yield satisfactory results, subject to a bit of 'tuning' at the commissioning stage. However, this state of affairs is unsatisfactory; to the designer, because he has no reliable means of evaluating the uncertainties in his calculations; to the component manufacturer, because he does not possess sufficiently complete knowledge to develop and optimise his designs to improve his competitive position; and to the acoustical engineer, because he does not feel confident in the validity of the assumptions and approximations employed, some of which seem not to be consistent with physical principles.

Experimental evaluation of the performance of HVAC components in the laboratory requires expensive, special-purpose ducts, so that the sound power generated by fans, or the dissipative attenuation of absorbers, can be determined from sound pressure measurements. It is impossible to make *in situ* measurements of the same quantities in operating systems; therefore, differences between laboratory and operational performance cannot be reliably established. Although laboratory generated data for behaviour of duct junctions and bends are available, there is no reliable evidence to show whether or not the application of these data to complex duct networks may be made with confidence. The development of means of making in-duct sound intensity measurements would largely remove these sources of uncertainty.

11.2   SOUND INTENSITY IN DUCTS WITHOUT MEAN FLOW

The analyses and figures of the sound intensity field in an infinite, uniform, two-dimensional duct, presented in Sections 3.7.3 and 4.8.4, reveal the complex nature of fields created by interference between the multiple reflection of sound waves from the duct boundaries; or, as an alternative physical view, the interaction between the pressure and particle velocity components of different duct modes, both propagating and non-propagating (evanescent). Intensity fields are particularly complex in the vicinity of sources of sound and duct discontinuities such as changes of cross-section, bends, junctions and terminations. As with the determination of sound power radiated by vibrating surfaces, the complex spatial distribution does not invalidate the application of Gauss's integral over any cross-section to the determination of the sound power flowing along a duct; it does, however, place greater demands on the intensity sampling procedure. Some theoretical examples are shown in Fig. 11.1.[167] As indicated in Chapter 4, intensity distributions in ducts at frequencies well above the lowest cut-off frequency are extremely complicated at single frequencies, but much simpler and more regular in form when generated by broad band sources and evaluated in frequency bands. As part of a study of the transmission of valve-generated noise through the walls of industrial pipework, measurements were made of the distribution of axial sound intensity over the cross-section of a large diameter pipe excited by a loudspeaker at frequencies up to eight times the lowest cut-off frequency f_{10}.[168] The distribution was found to be more uniform with the loudspeaker located off the pipe axis, and, with an open end to the pipe, the space-average axial intensity was found to be close to the spatial-average values of $\langle \overline{p^2} \rangle / \rho_0 c$ measured at the wall, and when averaged over the entire cross-section (Fig. 11.2). The acoustic pressures which excite a pipe wall can therefore be estimated from the sound power flowing down a pipe.

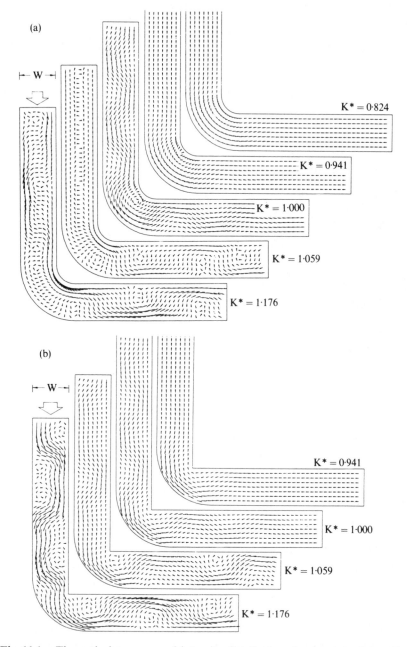

Fig. 11.1. Theoretical mean sound intensity distributions in a rectangular section, anechoically terminated duct having various bend configurations: $K^* =$ frequency/ cut-off frequency.[167]

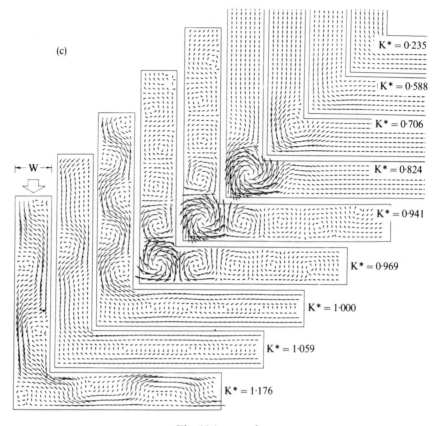

(c)

$K* = 0.235$

$K* = 0.588$

$K* = 0.706$

$K* = 0.824$

$K* = 0.941$

$K* = 0.969$

$K* = 1.000$

$K* = 1.059$

$K* = 1.176$

Fig. 11.1—*contd.*

Determinations have also been made of the division of sound power at junctions in ducts by measurements of sound intensity:[169,170] two examples are shown in Fig. 11.3(a, b). In the second example, estimates were made of the sound power flowing in a duct above the lowest cut-off frequency by terminating the duct in an expansion section, and scanning a probe over the outlet; this device also served to minimise adverse flow effects when measuring with airflow through the ducts.

Another HVAC application to which in-duct intensity measurement has been applied is the determination of the dissipative performance of various forms of attenuator. Figure 11.4 shows the effect of the form of duct termination on the *dissipative* attenuation of a simple 'reactive' expansion chamber device.[170] Clearly, more energy is dissipated if more is reflected from the termination. An implication of practical import is that the performance of attenuators measured in ducts having anechoic terminations may be poorer than that measured in ducts terminated by reverberation chambers, especially at low frequencies where the end reflection factor is large.

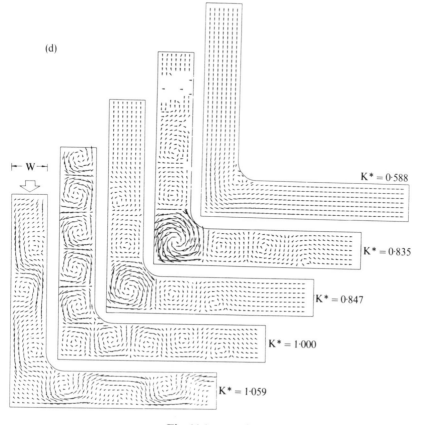

(d)

$K^* = 0.588$

$K^* = 0.835$

$K^* = 0.847$

$K^* = 1.000$

$K^* = 1.059$

Fig. 11.1—*contd.*

The second example of attenuator behaviour shows that a wooden plenum chamber, which is nominally a reactive attenuator, can dissipate energy to a significant degree, even at low frequencies (Fig. 11.5).[171]

11.3 SOUND INTENSITY IN STEADY MEAN FLOW

11.3.1 Sound intensity and energy density in ideal flow

In a homogeneous fluid in which the effects of viscosity and heat conduction may be assumed to be negligible, and in which the flow is irrotational (turbulence-free), one may identify two second-order energetic quantities which are related by a conservation equation analogous to eqn (4.11a). They are given by Refs 172 and 173 as

$$E = \tfrac{1}{2}(\rho_0|u|^2) + \tfrac{1}{2}(p^2/\rho_0 c^2) + (\mathbf{u}.\mathbf{U}/c^2)p \tag{11.1}$$

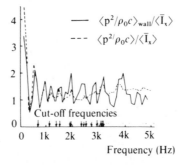

Fig. 11.2. Ratio of normalised mean square pressure to mean axial sound intensity in a duct of circular cross-section at frequencies extending well above the lowest cut-off frequency.[168]

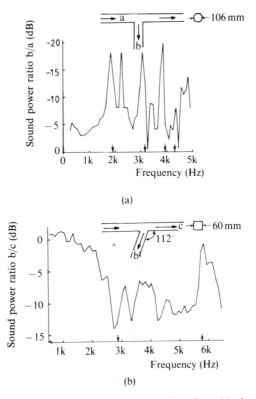

(a)

(b)

Fig. 11.3. Division of sound power at a junction in a duct: (a) circular duct;[168] (b) square duct.[169]

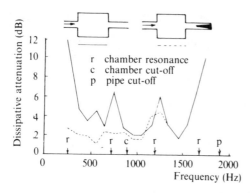

Fig. 11.4. Dissipative attenuation of a 'reactive' expansion chamber as a function of the duct termination.[170]

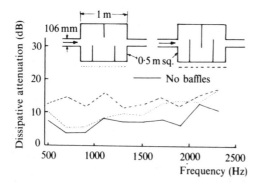

Fig. 11.5. Dissipative attenuation of a reactive plenum chamber.[171]

and

$$\mathbf{J} = (p/\rho_0 + \mathbf{u.U})(\rho_0\mathbf{u} + p\mathbf{U}/c^2) \qquad (11.2)$$

where **u** is the fluctuating (acoustic) component of the fluid velocity vector and **U** is the time-average (mean) component. E is a form of acoustic energy density and **J** is a form of acoustic intensity. The conservation equation linking them is

$$\partial E/\partial t + \nabla.\mathbf{J} = 0 \qquad (11.3)$$

In general two- or three-dimensional acoustic fields it is not possible to determine all the terms in eqn (11.2) from measurements of sound pressure at two closely spaced points, even when the mean flow is purely one-dimensional. This is made clear in a comprehensive analysis of the general three-dimensional intensity measurement problem by Munro and Ingard.[174]

11.3.2 Plane waves in one-dimensional mean flow

If the mean flow is one-dimensional, and the sound field takes the form of a plane *progressive* wave, the intensity may be expressed as

$$\mathbf{I} = (p^2/\rho_0 c)\{[1 + M(\mathbf{f.k})]\mathbf{k} + M[1 + M(\mathbf{f.k})]\mathbf{f}\} \tag{11.4}$$

in which the mean flow velocity vector is $U\mathbf{f}$, and the acoustic particle velocity vector is $u\mathbf{k}$. Three special cases may be identified:

(a) $\mathbf{k} = \mathbf{f}$: flow and sound propagation in the same direction; in which case

$$I = (p^2/\rho_0 c)(1 + M)^2 \tag{11.4a}$$

(b) Unit vectors $\mathbf{k}$ and $\mathbf{f}$ perpendicular; in which case
$$I = (p^2/\rho_0 c)[\mathbf{k} + M\mathbf{f}] \tag{11.4b}$$

(c) $\mathbf{k} = -\mathbf{f}$: sound propagation and flow in opposite directions; in which case

$$I = (p^2/\rho_0 c)(1 - M)^2 \tag{11.4c}$$

If the intensity were estimated from the usual p-p finite difference expression, the result corresponding to case (a) would be

$$I_0 = (p^2/\rho_0 c)/(1 - M) \tag{11.5a}$$

and that corresponding to case (c) would be

$$I_0 = (p^2/\rho_0 c)/(1 + M) \tag{11.5b}$$

In the general case of plane *progressive* waves, the fractional errors incurred by using the zero flow expressions for sound intensity in terms of the finite difference p-p approximation, or its spectral equivalent, are of the order of M, and therefore negligible for low speed flows, as also demonstrated by Chamant.[175]

11.3.3 Plane wave fields in ducts carrying steady mean flow

Where the direction of plane wave propagation is coincident with the mean flow direction, it is possible to express the sound intensity $\mathbf{J}$ in terms of the spectral and cross-spectral estimates of pressures measured at two closely spaced microphones, even in cases where reflections make the field partly reactive. Of course, this is subject to the validity of the assumptions made earlier of ideal, irrotational flow.[174] The expression for the magnitude of spectral component of intensity is

$$I(\omega) \approx [(1 - M^2)(1 + 3M^2)/\rho_0 \omega d]\,\text{Im}\,\{G_{12}\} + K[M(1 + M^2)/2\rho_0 c]$$
$$+ [Mc(1 - M^2)/\rho_0 \omega^2 d^2][(G_{11} - G_{22})^2 + 4(\text{Im}\,\{G_{12}\})^2]/K \tag{11.6}$$

where G_{11} and G_{22} are the spectral densities of the pressures at two microphones separated by a small distance d, G_{12} is their cross-spectral density, M is the mean flow Mach number and K is given by

$$K = G_{11} + G_{22} + 2\,\text{Re}\,\{G_{12}\}$$

This expression reduces to eqn (5.19) when $M = 0$, except for the different sign of the first term.

Jacobsen[173] simplified eqn (11.6) by neglecting terms of second order in M to give

$$I(\omega) \approx - \operatorname{Im}\{G_{12}\}/\rho_0 \omega d + (M/2\rho_0 c)(G_{11} + G_{22} + 2\operatorname{Re}\{G_{12}\})$$
$$+ (Mc/\rho_0 \omega^2 d^2)(G_{11} + G_{22} - 2\operatorname{Re}\{G_{12}\}) \tag{11.7}$$

In order to validate this expression, he performed experiments in a 10 cm diameter duct at flow speeds up to 30 m s^{-1}, and over a frequency range from 100 to 2000 Hz. He concluded that experimental evaluation of the third term in eqn (11.7) is subject to considerable uncertainty on account of its sensitivity to transducer phase and amplitude calibration and mismatch errors and the influence of uncorrelated noise. The magnitudes of the second and third (flow correction) terms depend upon position in an interference field, and are greatest at pressure maxima. Neglect of the flow correction terms can lead to gross errors in the estimate of intensity in highly reactive fields, even at very low Mach number. Errors of up to 10% in the estimate of Mach number can be tolerated. Warning was given of the serious problems created by unsteady flow phenomena which give rise to correlated 'noise' signals from the two pressure transducers. The small spacing necessary to effect the finite difference approximations to pressure and particle velocity can be considerably smaller than turbulence correlation scales, especially at low frequencies, and the resulting 'pseudo-intensity' then tends to swamp the true sound intensity. This problem is discussed further in Section 11.4.

It is also possible to formulate a general expression for sound intensity in plane, partly reactive, fields in ducts carrying flow in terms of relationships between the sound pressures measured at two points separated by an *arbitrary* axial distance. A number of different formulations have been published (e.g. Ref. 176). The following equation for net sound intensity in a uniform duct is based upon Ref. 177:

$$I(\omega) = \{(G_{11}/4\rho_0 c)\sin^2[ks/(1 - M^2)]\}$$
$$\times \{(1 + M)^2|\exp[jks/(1 - M)] - H_{12}|^2$$
$$- (1 - M)^2|H_{12} - \exp[-jks/(1 + M)]|^2\} \tag{11.8}$$

The expression is valid irrespective of the microphone separation distance s, provided that the flow is uniform and that energy dissipation between the two measurement points is negligible; H_{12} is the transfer function between the two microphone signals, one being considered as input and the other as output. The sensitivity of $I(\omega)$ to flow Mach number depends crucially on the reactivity of the acoustic field, through the relative magnitudes of the three terms under the modulus signs. (Note: eqn (11.8) differs from eqn (35) in Ref. 177—see Ref. 178.) It is clearly vital to ascertain M and c to a high degree of accuracy when applying eqn (11.8) to the analysis of experimental data (unlike the application of eqn (11.7)). Random errors in the estimate of H_{12} must also be minimised, by judicious selection of the microphone positions, and maximisation of the signal-to-noise ratio (if controllable).

Many other experimental techniques have been employed in order to extract the pressure amplitudes of the upstream- and downstream-propagating plane waves, primarily with the aim of determining impedances of duct components, and reflection coefficients of outlets (e.g. Ref. 179). Ref. 180 presents an analysis of the errors associated with two-microphone measurements in flow ducts.

11.4   SOUND INTENSITY IN UNSTEADY FLOW

11.4.1   Theory

Many attempts have been made to define acoustic energy density and energy flux quantities in unsteady flow of a viscous, heat conducting fluid. The fundamental difficulty is made clear by Morfey's analysis[172] of acoustic energy production (or dissipation) in such flows. The basic equation is

$$\partial E^*/\partial t + \partial N_i^*/\partial x_i = P \tag{11.9}$$

This is the acoustic energy equation, in which E^* represents a generalised acoustic energy density, and N_i^* are the components of a generalised energy flux. These quantities are given by

$$E^* = (\tfrac{1}{2}\rho c^2)(p'^2) + (V_j/c^2)(p'u_j) + \tfrac{1}{2}(\rho u_j^2) \tag{11.10}$$

and

$$N_i^* = (p'u_i) + (V_i/\rho c^2)(p'^2) + (V_i V_j/c^2)(p'u_j) + \rho V_j(u_i u_j) \tag{11.11}$$

in which p' is the fluctuating component of pressure, V is the total particle velocity and $u = v - w$, where v is the total fluctuating velocity and w is the fluctuating velocity associated with the vorticity of the flow: tensor suffix notation applies. The quantity P is the rate of acoustic energy production per unit volume associated with viscous and heat conduction effects, and interactions of sound with flow; it may be negative and represent net dissipation, or positive and represent sound generation, and it vanishes in irrotational, uniform entropy flows. Note that p' is associated with both compressible and vortical field components. The identification of the quantity which is of practical significance in relation to the engineering requirements outlined in the introduction to this chapter represents a major theoretical difficulty.

11.4.2   Measurements of 'sound intensity' in unsteady flow

There are three main obstacles to the use of the two-microphone intensity measurement technique in turbulent fluid flow. Even in steady, irrotational flow, the presence of higher order, transverse duct modes renders the output from a two-microphone probe ambiguous.[174] In rotational, turbulent flow, it is, in principle, impossible to separate the fluctuating particle velocities associated with incompressible unsteady flow from those associated with the compressible acoustic field. Finally, turbulence transported by the flow, or generated by the presence of the probe, produces pressure fluctuations on the microphones, which are falsely interpreted as acoustic signals.

In the face of the theoretical impediments, the author has pursued an essentially empirical approach to the problem of measurement of acoustic energy flux in low speed, turbulent airflow.[181] A face-to-face half-inch condenser microphone pair has been enveloped in a 50 mm diameter, cigar-shaped windscreen (Fig. 11.6). The principle of discrimination against turbulence is illustrated qualitatively by Fig. 11.7. The turbulent pressures acting on the outer surface of the windscreen generate a through-screen fluctuating velocity field, just as do the sound field pressures (the screen cannot discriminate). However, the turbulence-driven field contains only axial (flow direction) wavenumbers greater than k; hence it generates only a sound field which

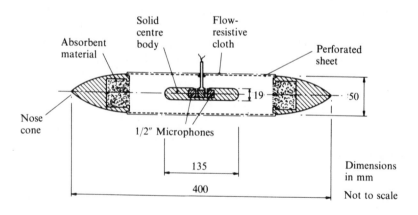

Windscreened probe

Fig. 11.6. ISVR windscreened sound intensity probe.

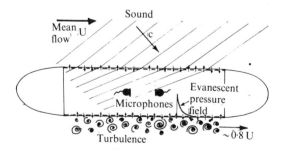

Fig. 11.7. Illustration of the principle of a windscreen.

decays exponentially with distance from the screen, the decay exponent being closely proportional to frequency. On the other hand, the sound-driven velocity field propagates freely into the enclosed volume. Hence, the larger the radius of the windscreen, the greater the discrimination between sound and turbulence. This probe was used in a wind tunnel carrying airflow at speeds up to $27\,\mathrm{m\,s^{-1}}$, and in a frequency range which included four transverse modes. Good agreement was obtained in the frequency range 315–5000 Hz between an intensity traverse of the working section and a traverse with a standard intensity probe over a measurement surface in virtually quiescent air enclosing the wind tunnel intake (Fig. 11.8). No correction for flow was applied to the cross-spectral estimate of intensity, since no theoretical correction is possible at frequencies above the lowest cut-off frequency f_{10}.

Subsequent tests in a commercial ventilation fan test duct (courtesy of Sound Attenuators Ltd, Colchester, UK) showed that rejection of large-scale turbulence below 250 Hz was completely inadequate; intense, highly correlated turbulence signals from each microphone produced intensity estimates up to 21 dB greater than the true values. In order to generate a signal of which the turbulent component was uncorrelated with those from the windscreened microphones, but of which the acoustic component was highly correlated with the latter, a third microphone in a turbulence suppression tube was employed.[182] On the basis of the assumption of perfect acoustic correlation, a corrected estimate of the cross-spectral density between the wind-screened probe microphones is

$$G_{12} = G_{14}G_{24}^{*}/(\gamma_{14}^{2}G_{11}G_{44})^{1/2} \tag{11.12}$$

where microphones 1 and 2 are in the probe, microphone 4 is in the turbulence suppression tube, and γ^{2} is the coherence function: in this equation it is assumed that the sensitivities of the probe and tube microphones are the same. A comparison of corrected and uncorrected intensity estimates is presented in Fig. 11.9.

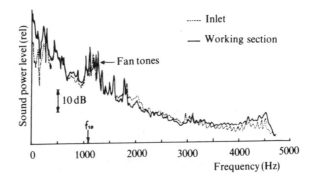

Fig. 11.8. Comparison between sound power estimated in a wind tunnel test section at $27\,\mathrm{m\,s^{-1}}$ and at the intake.[181]

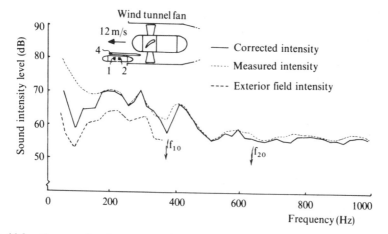

Fig. 11.9. Corrected and uncorrected estimates of sound intensity made in the highly turbulent wake of a fan at $12\,\mathrm{m\,s^{-1}}$ [182]

A very instructive experimental comparison has been made between the capacities for turbulence suppression tubes, streamlined microphone nose cones and windscreened intensity probes to suppress turbulence-generated disturbances in mean air speeds of up to $30\,\mathrm{m\,s^{-1}}$.[183] The conclusion was that, in the low frequency range (25–200 Hz), in moderate turbulence, the intensity probe in the pressure mode and the turbulence suppression tube are similarly effective, but that the former is superior at higher frequencies. The intensity probe in the intensity mode was superior to both above about 630 Hz. It should be emphasised that the adverse influence of turbulence on intensity measurement depends not only on the turbulence intensity, but strongly on its spatial scale, thereby constituting a more serious problem in ducts of large cross-sectional area, typical of low speed ventilation ducts, than in smaller ducts typical of laboratory test arrangements. A recent experimental study of the effect of moderate airflow on sound intensity measurement is presented in Ref. 61.

11.5   MEASUREMENTS OF SOUND INTENSITY IN LIQUID-FILLED PIPES

The principle of measurement of sound intensity in liquid-filled pipes is no different from that in gases. However, in pipes typical of hydraulic power systems, or liquid transport ducts, the lowest cut-off frequency f_{10} is usually in the order of thousands of Hz, and one-dimensional wave propagation is dominant. Also, pipe walls stretch in response to internal fluid pressure, and vibrational power flow in the walls accompanies acoustic power flow in the fluid; indeed, the flexibility of the walls profoundly affects the wave propagation characteristics.[184] The effect on wave phase speed can be significant when attempting to apply forms of expression for intensity such as eqn (11.8), in which k must be known accurately.

A number of experimental investigations have been carried out on pipework installations, including those by Badie-Cassagnet *et al.*[185] and Verheij.[186] The former employed pairs of pressure transducers in the wall of an industrial hydraulic circuit, and clearly demonstrated the dissipative attenuation of a line filter (Fig. 11.10). The latter inferred vibrational power flow in a pipe from measurements of the wall acceleration with arrays of accelerometers.

Chamant[175] developed a streamlined, multi-transducer intensity probe with which he investigated sound propagation in a water tunnel, and also the radiation of sound from the propeller of a boat under way at $5\,m\,s^{-1}$. The conclusion from this limited study was that the intensity probe was superior to

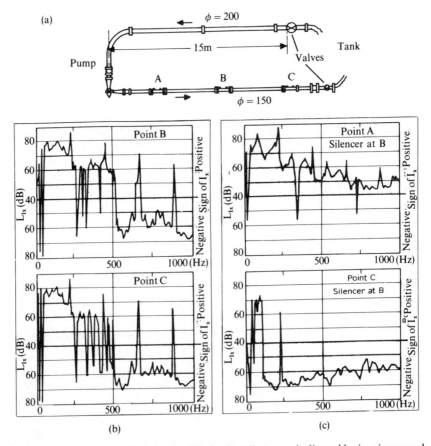

Fig. 11.10. Effectiveness of a hydraulic pipeline silencer as indicated by in-pipe sound intensity measurements: (a) the circuit; (b) intensities at points A and C without a silencer; (c) intensities at points A and C with a silencer at B.[185]

a single hydrophone in terms of discrimination of sound from turbulence at frequencies in excess of 250 Hz, but that it was excessively sensitive to unsteady hydrodynamic pressure at lower frequencies.

11.6 CONCLUSIONS

Attempts to develop methods of measurement of sound power flux in moving, turbulent fluids lie, at the time of writing, firmly in the realms of research. It seems highly unlikely that measurements in airflow at speeds over $50 \, \text{m s}^{-1}$ will ever be feasible, mainly because of the noise generated by the presence of any structure designed to screen the pressure sensors from oncoming turbulence.

Intensity standards

INTERNATIONAL STANDARDS

ISO 9614-1 Acoustics—Determination of sound power levels of noise sources by sound intensity measurement. Part 1—Measurement at discrete points

ISO 9614-2 Acoustics—Determination of sound power levels of noise sources by sound intensity measurement. Part 2—Measurement by scanning

IEC 1043 Instruments for the measurement of sound intensity

NATIONAL STANDARDS

USA

ANSI S.12.12—1992 Engineering method for the determination of sound power level of noise sources using sound intensity

France

NF EN 29614-1 Détermination par intensimétrie des niveaux de puissance acoustique émis par les sources de bruit—Partie 1: mesurages par points

NF EN 61043 Instruments pour la mesure de l'intensité

United Kingdom

BS ISO 9614-1 Acoustics—Determination of sound power levels of noise sources by sound intensity measurement. Part 1—Measurement at discrete points

BS ISO 9614-2 Acoustics—Determination of sound power levels of noise sources by sound intensity measurement. Part 2—Measurement by scanning

Czech Republic and Slovakia

CSN ISO 9614-1 Determination of sound power levels of noise sources using sound intensity measurement—Measurement at discrete points

Norway

NS—INSTA 121 Acoustics. Determination of sound power levels of noise sources using sound intensity measurements. Scanning method for use *in situ*

Sweden

SS 459 01 10 Identical to NS—INSTA 121

EUROPEAN COMPUTER MANUFACTURERS' ASSOCIATION STANDARD

ECMA—160 Determination of sound power levels of computer and business equipment using sound intensity measurements; scanning method in controlled rooms

Notation

a	Width of duct; radius of neck of Helmholtz resonator
A	Complex amplitude of a harmonically varying quantity
$\mathbf{A}$	General vector quantity
A_m	Measured average
A_n	Component of $\mathbf{A}$ in direction $\mathbf{n}$
A_t	True average
B	Complex amplitude of a harmonically varying quantity; filter bandwidth (Hz); ϕ_s/kd
c	Speed of propagation of unconvected ultrasonic beam and speed of propagation of small acoustic disturbances in a fluid (speed of sound)
c_g	Wave group speed
C	Complex intensity of a harmonic sound field; numerical factor
C_{12}	Real part of G_{12}
d	Distance between acoustic centres of p-p intensity probe transducers; distance between sender and receiver in p-u intensity probe
d/dx	Derivative with respect to x
D/Dt	Total time derivative
e	Total acoustic energy density in non-flowing fluid
$e(\,)$	Normalised systematic (bias) error caused by finite difference approximation inherent in p-p measurements and by finite ultrasonic beam length in p-u probes
$e_b(\,)$	Normalised systematic (bias) error
e_p, e_k	Potential and kinetic acoustic energy densities
$e_r(\,)$	Normalised random error
$e_\phi(\,)$	Normalised systematic error caused by transducer/analyser channel phase mismatch
E	Vibrational energy of a plate; acoustic energy density in ideal steady flow
E^*	Generalised acoustic energy density in unsteady flow
$\mathbf{f}$	Vector force per unit volume; unit normal in mean flow direction
f	Frequency (Hz); arbitrary wave function
f_i	Component of $\mathbf{f}$ in i direction
$\mathbf{F}$	Force vector

$\left.\begin{array}{l} F_1 \\ F_2 \\ F_3 \\ F_4 \\ F_{pl} \\ F_{\pm} \end{array}\right\}$ Sound field indicators

g — Arbitrary wave function

G — Harmonic free-space Green function

$G_{xx}(\omega)$ — One-sided spectral density of quantity (signal) x $(x^2/\text{rad s}^{-1})$

$G_{xy}(\omega)$ — One-sided cross-spectral density of quantities (signals) x and y

G'_{xy} — Corrected cross-spectral density

h — $d/2$

H_x — Sensitivity of transducer x

H_{12} — Transfer function between input 1 and output 2

i — Square root of -1

$\mathbf{i}$ — Unit vector defining direction of x co-ordinate axis

I — Magnitude of the mean (time-average) intensity $(=\bar{I}_a)$; real (active) component of complex intensity C

$\mathbf{I}$ — Mean intensity vector

$I(t)$ — Magnitude of the instantaneous intensity vector

$\mathbf{I}(t)$ — Instantaneous intensity vector

I_a — Active component of intensity in a harmonic field

I_b — Value of I in a frequency band

I_e — Estimated value of I

I_i — Indicated value of I

I_n — Component of $\mathbf{I}$ in direction $\mathbf{n}$; normal component of $\mathbf{I}$

$I_n(t)$ — Magnitude of the component of $\mathbf{I}(t)$ in direction $\mathbf{n}$

I_{nx} — Axial component of $\mathbf{I}$ in isolated acoustic duct mode n

I_r — Component of mean intensity in direction $\mathbf{r}$: radial component of I

I_{re} — Reactive component of intensity in a harmonic field

I_x, I_y — x- and y-directed components of I

I_0 — Reference value of I in a plane progressive wave

I_θ — Tangential component of $\mathbf{I}$

Im { } — Imaginary part of

$\mathbf{j}$ — Unit vector defining direction of y co-ordinate axis

$\mathbf{J}$ — Reactive intensity vector: acoustic intensity in ideal steady flow (Chapter 11)

$\mathbf{k}$ — Unit vector defining direction of z co-ordinate axis; wave vector

k — Wavenumber: magnitude of $\mathbf{k}$

k_i — Wavenumber component of k in direction i

k_n, k_m — transverse wavenumbers of acoustic modes m, n $(= m\pi/a, n\pi/a)$

k_t — Surface wavenumber

K — Non-dimensional wavenumber ratio k/k_t; bias error factor

l' — Effective length of neck of Helmholtz resonator

L	Lagrangian $[e_k - e_p]$
L_b	Frequency band level of random error due to random error in estimate of phase mismatch (or its correction)
L_d	Dynamic capability index
$L_{|Ii|}$	Indicated sound intensity level
L_{In}	Normal sound intensity component level
L_p	Sound pressure level
L_r	Frequency band level of random error due to random errors in spectral line estimates
L_w	Sound power level
L_ϕ	Phase error index
m, n	Acoustic duct mode orders
M	Mach number; dipole strength
$\mathbf{n}$	Unit vector; unit normal vector
n	Number of spectral lines in a frequency band; integer
N	Integer
N_i^*	Component of generalised energy flux
p	Acoustic (sound) pressure
p'	Fluctuating fluid pressure
p^+, p^-	Acoustic (sound) pressures in plane progressive waves travelling in the positive and negative x-directions
$\overline{p^2}$	Mean square acoustic (sound) pressure
p_e	Estimated value of p
p_{ref}	$20\,\mu\text{Pa m}^{-2}$
p-p	Intensity probe comprising two nominally identical pressure transducers
p-u	Intensity probe comprising a pressure transducer and a particle velocity transducer
P	Total fluid pressure; complex amplitude of harmonic sound pressure
P_0	Mean (static) fluid pressure
P_g	Fluid gauge pressure
P_m, P_n	Pressure amplitudes of acoustic duct modes m, n
P_{1r}, P_{2r}	Real components of sound pressures of sources 1, 2
P_{1i}, P_{2i}	Imaginary components of sound pressures of sources 1, 2
q	Volumetric velocity source (strength) density
Q	Volumetric velocity (strength) of monopole source; dynamic magnification factor of Helmholtz resonator
Q'	Volumetric velocity per unit length of a line monopole source
Q_{12}	Imaginary part of G_{12}
r	Position vector; radial spherical co-ordinate
R	Magnitude of complex sound pressure reflection factor; internal resistance of Helmholtz resonator; sound reduction index
R_r	Radiation resistance of Helmholtz resonator
$R_{xy}(\tau)$	Cross-correlation function for quantities (signals) x, y at time delay τ

$\text{Re}\{\}$	Real part of
s	Path length co-ordinate; normalised density change (condensation)
S	Area of measurement surface; area of enclosure surface; area of neck of Helmholtz resonator
$S_{xx}(\omega)$	Two-sided spectral density of quantity (signal) $x(x^2/\text{rad s}^{-1})$
$S_{xy}(\omega)$	Two-sided cross-spectral density of x and y
t	time; Student's t factor; temperature (°C)
t^+, t^-	Incremental beam transit times in the positive and negative axial directions
T	Integration period; averaging time
T_{60}	Reverberation time
T^+, T^-	Ultrasonic beam transit times between transducers
TL	Sound transmission loss
$\mathbf{u}$	Acoustic particle velocity vector
u	Component of $\mathbf{u}$ in the x-direction; convection speed
u^+, u^-	Particle velocities in plane progressive waves travelling in the positive and negative x-directions
u'	Fluctuating particle velocity
u_n	Normal component of acoustic particle velocity
u_r	Radial component of $\mathbf{u}$
u_θ, u_ϕ	Tangential components of $\mathbf{u}$
U	Complex amplitude of x-directed component of harmonic particle velocity; potential energy density; mean fluid velocity
U_e	Estimated value of complex amplitude of particle velocity
U_r	Complex amplitude of radial component of harmonic particle velocity
U_ri	Imaginary part of U_r
U_rr	Real part of U_r
U_θ	Complex amplitude of tangential component of harmonic particle velocity
$U_{\theta\mathrm{i}}$	Imaginary part of U_θ
$U_{\theta\mathrm{r}}$	Real part of U_θ
v	Component of $\mathbf{u}$ in the y-direction; total fluctuating velocity component
v_n	Surface normal velocity
v^*	Normalised standard deviation
V	Volume; complex amplitude of y-directed component of harmonic particle velocity
V_i	component of total fluid velocity
V_n	Complex amplitude of harmonic normal velocity of a surface
w	Component of $\mathbf{u}$ in the z-direction; vortical component of velocity
W	Work; radiated power; complex amplitude of z-directed component of harmonic particle velocity
W'	Rate of work (power per unit volume)
W_{nx}	Sound power propagated by acoustic duct mode n
W_s	Time-average sound power of all sources within surface S

W_t	Sound power transmitted through a partition
x	Cartesian co-ordinate
y	Cartesian co-ordinate
z	Cartesian co-ordinate; specific acoustic impedance
α	Sound absorption coefficient; phase angle (Fig. 3.5)
β	Phase angle (Fig. 3.5)
γ	Ratio of specific heats
γ_{12}^2	Coherence function between quantities (signals) 1 and 2
δx	Small increment in x
δ_{pI}	Pressure-intensity index
δ_{pI0}	Pressure-residual intensity index
$\partial/\partial x$	Partial derivative with respect to x
η	y-directed component of particle displacement vector
ζ	z-directed component of particle displacement vector
θ	Spherical co-ordinate; phase of complex sound pressure reflection-factor
λ	Wavelength
ρ	Fluid density
ρ_0	Mean fluid density
ϕ	Spherical co-ordinate
ϕ_f	Field phase difference
ϕ_m	Angle of deviation of the direction of minimum p-p intensity sensitivity from the ideal ($\pi/2$)
ϕ_p	Phase of pressure
ϕ_r	Relative phase between pressure and particle velocity
ϕ_s	Phase mismatch of probe/analyser chain
ϕ_u	Phase of particle velocity
ϕ_0	Plane progressive wave p-p probe phase difference (kd)
σ	Standard deviation; radiation efficiency
τ	Time delay; sound power transmission coefficient
$\boldsymbol{\xi}$	Particle displacement vector
ξ	x-directed component of $\boldsymbol{\xi}$
ω	Circular frequency (rad s^{-1})
ω_n	Frequency of spectral line
ω_u	Frequency of ultrasonic beam
ω_0	Natural frequency of Helmholtz resonator

MATHEMATICAL OPERATIONS

$*$	Complex conjugate
$\| \ \|$	Modulus of
$\langle \ \rangle$	Space-average
—	Time-average
$'$	First partial space derivative

$''$	Second partial space derivative
$'''$	Third partial space derivative
iv	Fourth partial space derivative
n	nth partial space derivative
·	Scalar product
×	Vector product
$\sum$	Summation
$\nabla\cdot$	Divergence
$\nabla\times$	Curl (rotation)
$F\{\}$	Fourier transform
$\wedge$	Hilbert transform

References

1. Wolff, I. and Massa, F., Direct measurement of sound energy density and sound energy flux in a complex sound field. *J. Acoust. Soc. Amer.*, **3** (1932) 317–18.
2. Olson, H. F., Field-type acoustic wattmeter. *J. Audio Engng. Soc.*, **22** (1974) 321–8.
3. Enns, J. H. and Firestone, F. A., Sound power density fields. *J. Acoust. Soc. Amer.*, **14** (1942) 24–31.
4. Clapp, C. W. and Firestone, F. A., The acoustic wattmeter, an instrument for measuring sound energy flow. *J. Acoust. Soc. Amer.*, **13** (1941) 124–36.
5. Bolt, R. H. and Petrauskas, A. A., An acoustic impedance meter for rapid field measurements. *J. Acoust. Soc. Amer.*, **15** (1943) 79(a).
6. Baker, S., Acoustic intensity meter. *J. Acoust. Soc. Amer.*, **27** (1955) 269–73.
7. Schultz, T. J., Acoustic wattmeter. *J. Acoust. Soc. Amer.*, **28** (1956) 693–9.
8. Schultz, T. J., Smith, P. W. and Malme, C. I., Measurement of sound intensity in a reactive sound field. *J. Acoust. Soc. Amer.*, **57** (1975) 1263–8.
9. Stenzel, H., *Leitfaden zur Berechnung der Schallvorgänge*, Berlin, 1958.
10. Morse, P. M., *Vibration and Sound*, 2nd edn. McGraw-Hill, New York, 1948.
11. Mechel, F., Neues Verfahren zur Messung von Wandimpedanzen und Wirkleistungen. *Verhandl DPG(VI)*, **3** (1968) 391.
12. Odin, G., Beitrag zur Messung der Schallintensität. In *Proceedings of IV Akust. Konferenz*, Budapest, 1967.
13. Kurze, U., Zur Entwicklung eines Gerätes für komplexe Schallfeldmessungen. *Acustica*, **20** (1968) 308–10.
14. Burger, J. F., van der Merwe, G. J. J., van Zyl, B. G. and Joffe, L., Measurement of sound intensity applied to the determination of radiated sound power. *J. Acoust. Soc. Amer.*, **53** (1973) 1167–8.
15. Van Zyl, B. G. and Anderson, F., Evaluation of the intensity method of sound power determination. *J. Acoust. Soc. Amer.*, **57** (1975) 682–6.
16. Van Zyl, B. G., Council for Scientific and Industrial Research (CSIR), Pretoria, Republic of South Africa. Sound Intensity Meter. *Technical Information for Industry*, **17** (1979) 1–4.
17. Van Zyl, B. G., Bepaling van klankdrijwing met behulp van een klankintensiteitsmeter. Master thesis, University of Pretoria, Republic of South Africa, 1974.
18. Lambrich, H. P. and Stahel, W. A., A sound intensity meter and its applications in car acoustics. In *Proceedings of Inter-Noise 77*, ed. Eric J. Rathe. International Institute of Noise Control Engineering, 8332 Zurich-Russihon, Switzerland.
19. Pavic, G., Measurement of sound intensity. *J. Sound Vib.*, **51** (1977) 533–46.
20. Fahy, F. J., A technique for measuring sound intensity with a sound level meter. *Noise Control Engng*, **9** (1977) 155–62.
21. Intensimètre Acoustique INAC 201. Metravib, 64 chemin des Mouilles, BP 182, 69132 Ecully cedex, France.
22. Roth, O., A sound intensity real-time analyser. In *Proceedings of Recent Developments in Acoustic Intensity Measurement*, ed. M. Bockhoff. Centre Technique des Industries Mécaniques, Senlis, France, 1981, pp. 69–74.

23. Miller, A. S., An investigation of the measurement of acoustic intensity using digital signal processing. BSc Dissertation, Department of Mechanical Engineering, University of Southampton, 1976.

24. Alfredson, R. J., A new technique for noise source identification on a multi-cylinder automotive engine. In *Proceedings of Noise-Con 77*, ed. George C. Maling. Noise Control Foundation, New York, 1977, pp. 307–18.

25. Lambert, J. M. and Badie-Cassagnet, A., La mesure directe de l'intensité acoustique. CETIM-Informations No. 53, Centre Technique des Industries Mécaniques, Senlis, France, 1977, pp. 78–97.

26. Chung, J. Y., *Cross-spectral Method of Measuring Acoustic Intensity*. Research Publication, General Motors Research Laboratory, GMR-2617, Warren, Michigan, 1977.

27. Chung, J. Y., Cross-spectral method of measuring acoustic intensity without error caused by instrument phase mismatch. *J. Acoust. Soc. Amer.*, **64** (1978) 1613–16.

28. Fahy, F. J., Measurement of acoustic intensity using the cross-spectral density of two microphone signals. *J. Acoust. Soc. Amer.*, **62** (L) (1977) 1057–9.

29. Fahy, F. J., A technique for measuring sound intensity with a sound level meter. In *Proceedings of the Ninth International Congress on Acoustics*, ed. Anon. Spanish Acoustical Society, c/o Serrano 144, Madrid (1977) p. 824.

30. Nordby, S. A. and Bjor, O-H., Measurement of sound intensity by use of a dual channel real-time analyser and a special sound intensity microphone. In *Proceedings of Inter-Noise 84*, ed. George C. Maling. Noise Control Foundation, New York, USA, 1984, pp. 1107–10.

31. Bjor, O.-H. and Krystad, H. J., A velocity microphone for sound intensity measurement. In *Proceedings of the Autumn Conference 1982*, ed. Anon. Institute of Acoustics, Edinburgh, 1982, pp. B7.1–B7.5.

32. Kinsler, L. E., Frey, A. R., Coppens, A. B. and Sanders, J. V., *Fundamentals of Acoustics*, 3rd edn. John Wiley, New York, 1982.

33. Fahy, F. J., *Sound and Structural Vibration*. Academic Press, London, 1987.

34. Blake, W. K., *Mechanics of Flow-Induced Sound and Vibration*. Academic Press, Orlando, USA, 1986.

35. Lighthill, M. J., *Waves in Fluids*. Cambridge University Press, Cambridge, 1978, p. 15.

36. Pascal, J. C., Structure and patterns of acoustic intensity fields. In *Proceedings of the Second International Congress on Acoustic Intensity*, ed. M. Bockhoff, Centre Technique des Industries Mécaniques, Senlis, France, 1985, pp. 97–104.

37. Adin Mann, J., III, Tichy, J. and Romano, A. J., Instantaneous and time-averaged energy transfer in acoustic fields. *J. Acoust. Soc. Amer.*, **82** (1987) 17–30.

38. Uosukainen, S., Properties of acoustic energy quantities. *Research Report 656*, Technical Research Centre of Finland, Espoo, Finland, 1989.

39. Zhong, Q. and Alfredson, R. J., The instantaneous sound intensity in two-dimensional sound fields—a finite element approach. *Acoustics Australia*, **21** (1993) 22–7.

40. Jacobsen, F., A note on instantaneous and time-averaged active and reactive intensity. *J. Sound Vib.*, **147** (1991) 489–96.

41. Heyser, R. C., Instantaneous sound intensity. In *Proceedings of the 81st Convention of the Audio Engineering Society, Preprint 2399*, 1986.

42. Schultz, T. J., Persisting questions in steady-state measurements of noise power and sound absorption. *J. Acoust. Soc. Amer.*, **54** (1973) 978–84.

43. Jacobsen, F., The diffuse sound field. *Report No. 27*, The Acoustics Laboratory, Technical University of Denmark, Lyngby, Denmark, 1979.

44. Ebeling, K. J., Statistical properties of random wave fields. In *Physical Acoustics*, Vol. XVII, eds W. P. Mason, and R. N. Thurston, Academic Press, New York, 1984.

45. Jacobsen, F., Active and reactive sound intensity in a reverberant sound field. *J. Sound Vib.*, **143** (1990) 231–40.
46. Waterhouse, R. V., Energy streamlines for an extended source. In *Proceedings of the Second International Congress on Acoustic Intensity*, ed. M. Bockhoff, Centre Technique des Industries Mécaniques, Senlis, France, 1985, pp. 129–35.
47. Kristiansen, U. R., Numerical experiments on sound radiation from beams. In *Proceedings of the Eleventh International Congress on Acoustics*, ed. P. Lienard, Groupements des Acousticiens de Langue Français, Paris, 1983, pp. 99–102.
48. Waterhouse, R. V., Crighton, D. G. and Ffowcs-Williams, J.E., A criterion for an energy vortex in a sound field. *J. Acoust. Soc. Amer.*, **81** (1987) 1323–6.
49. Fahy, F. J. and Elliott, S. J., Acoustic intensity measurements of transient noise sources. *Noise Control Engng*, **17** (1981) 120–5.
50. Arai, M., Correlation methods for estimating radiated power. In *Proceedings of the Sixth International Congress on Acoustics*, ed. Y. Kohasi, Mazuren Co. Ltd, Tokyo, 1968, pp. F-13–F-16.
51. Macadam, J. A., The measurement of sound radiation from room surfaces in lightweight buildings. *Applied Acoustics*, **9** (1976) 103–18.
52. Brito, J. D., Sound intensity patterns for vibrating surfaces. PhD thesis, Massachusetts Institute of Technology, Cambridge, Mass. 1976.
53. Hodgson, T. H. and Chun, Du H., Development of a surface acoustic intensity probe. In *Proceedings of Inter-Noise 84*, ed. George C. Maling. Noise Control Foundation, New York, 1984, pp. 1087–92.
54. Anon., Characteristics of microphone pairs and probes for sound intensity measurement. Bruel and Kjaer report, Naerum, Denmark, 1987.
55. Rebillat, J. C. and Rifai, S., Détermination expérimentale de la diffraction acoustique autour de sondes intensimétriques. *Acustica*, **70** (1990) 83–8.
56. Fredriksen, E. and Schultz, O., Pressure microphones for intensity measurement with significantly improved phase properties. In *Bruel and Kjaer Technical Review No 4–1986*. Bruel and Kjaer, Naerum, Denmark, 1986, pp. 11–23.
57. Zaveri, K., Influence of tripods and microphone clips on the frequency response of microphones. In *Bruel & Kjaer Technical Review No 4–1985*. Bruel and Kjaer, Naerum, Denmark, 1985, pp. 32–40.
58. Kuttruff, H. and Schmitz, A., Measurement of sound intensity by means of multi-microphone probes. *Acustica*, **80** (1994) 388–96.
59. Kuttruff, H., Messung der Schallintensität mit einer Vielfalch-Mikrophonanordung. *Acustica*, **72** (1990) 161–5.
60. Pascal, J. C. and Carles, C., Systematic measurement errors with two microphone sound intensity meters. *J. Sound Vib.*, **83** (1982) 53–65.
61. Jacobsen, F., Intensity measurements in the presence of moderate airflow. In *Proceedings of Inter-Noise 94*, edited by S. Kuwano, Institute of Noise Control Engineering, Japan, 1994, pp. 1737–1742.
62. Jacobsen, F., A note on the measurement of sound intensity with windscreened probes. *Applied Acoustics*, **42** (1994) 41–53.
63. Jacobsen, F. and Olsen, E. S., The influence of microphone vents on the performance of sound intensity probes. *Applied Acoustics*, **41** (1993) 25–45.
64. Bronsdon, R. and Hargrave, R., Tape recording digitized acoustic intensity data using a video tape recorder. In *Proceedings of Inter Noise 84*, ed. George C. Maling. Noise Control Foundation, New York, 1984, p. 1051–4.
65. Watkinson, P. S., Techniques and instrumentation for the measurement of transient sound energy flux. *Technical Report No 122*, Institute of Sound and Vibration Research, University of Southampton, 1983.
66. Jarvis, D. R., The calibration of sound intensity instruments. *Acustica*, **80** (1994) 103–14.

67. Bodén, H. and Åbom, M., Influence of errors on the two-microphone method for measuring acoustic properties of ducts. *J. Acoust. Soc. Amer.*, **79** (1986) 541–9.

68. Fredriksen, E., Acoustic calibrator for intensity measurement systems. In *Bruel and Kjaer Technical Review No 4–1987*. Bruel and Kjaer, Naerum, Denmark, 1987, pp. 36–42.

69. Jacobsen, F., A note on the accuracy of phase-compensated intensity measurements. *J. Sound Vib.*, **174** (1994) 140–4.

70. Krishnappa, G., Cross spectral method of measuring acoustic intensity by correcting phase and gain mismatch errors by microphone calibration. *J. Acoust. Soc. Amer.*, **69** (1981) 307–10.

71. Patrat, J.-C., Intensimétrie acoustique à l'aide de matériel non-spécialisé: problèmes d'étalonage. In *Proceedings of the Second International Congress on Acoustic Intensity*, ed. M. Bockhoff, Centre Technique des Industries Mécaniques, Senlis, France, 1985, pp. 31–8.

72. Jacobsen, F., A simple and effective correction for phase mis-match in intensity probes. *Applied Acoustics*, **33** (1991) 165–80.

73. Ren, M. and Jacobsen, F., A simple technique for improving the performance of intensity probes. *Noise Control Engng. Jour.*, **38** (1992) 17–25.

74. Jacobsen, F., Random errors in sound intensity estimation. *J. Sound Vib.*, **128** (1989) 247–57.

75. Herlufsen, H., Dual Channel FFT Analysis (Part 1). *Bruel and Kjaer Technical Review No 1–1984*. Bruel and Kjaer, Naerum, Denmark, 1984.

76. Pascal, J. C., Analytical expressions for random errors of acoustic intensity. In *Proceedings of Inter-noise 86*, ed. Robert Lotz. Noise Control Foundation, New York, 1986, pp. 1073–6.

77. Seybert, A. F., Statistical errors in acoustic intensity measurements. *J. Sound Vib.*, **75** (1981) 519–26.

78. Jacobsen, F. and Nielson, T., Spatial correlation and coherence in a reverberant sound field. *J. Sound Vib.*, **118** (1987) 175–80.

79. Jacobsen, F., Active and reactive, coherent and incoherent sound fields. *J. Sound Vib.*, **130** (1989) 493–507.

80. Cook R. K., A new method for the measurement of total energy density of sound waves. In *Proceedings of Inter-Noise 74*, ed. J. C. Snowdon, Institute of Noise Control Engineering, P.O. Box 1758, Poughkeepsie, NY 12601, USA, pp. 101–6.

81. Riedlinger, R., Un appareil de mesure de la densité de l'énergie dans l'air. In *Proceedings of the 2nd International Congress on Acoustic Intensity*, ed. M. Bockhoff, CETIM, F-60304 Senlis, France, pp. 59–68.

82. Pesonen, K. and Uosukainen, S., On rotationality of acoustic intensity fields. In *Proceedings of Inter-Noise 84*, ed. George C. Maling. Noise Control Foundation, New York, 1984, pp. 1125–8.

83. Kihlman, T. and Tichy, J., Studies of sound intensity measurements on a steel panel. In *Proceedings of Inter-Noise 87*, ed. Li Peizi. Acoustical Society of China, P.O. Box 2712, Beijing, People's Republic of China, 1987, pp. 1231–4.

84. Hübner, G., Determination of sound power of sources under in-situ conditions using intensity method—field application, suppression of parasitic noise, reflection effect. In *Proceedings of Inter-Noise 83*, ed. R. Lawrence. Institute of Acoustics, Edinburgh, 1983, pp. 1043–6.

85. Laville, F., The optimal measurement distances for intensity determination of sound power. In *Proceedings of Inter-Noise 83*, ed. R. Lawrence. Institute of Acoustics, Edinburgh, 1983, pp. 1075–8.

86. Jacobsen, F., Sound field indicators: useful tools. *Noise Control Engng Jour.*, **35** (1990) 37–46.

87. Nelson, P. A., Personal communication.
88. Vigran, T. E., Acoustic intensity–energy density ratio: an index for detecting deviations from ideal field conditions. *J. Sound Vib.*, **127** (1988) 343–51.
89. Ren, M., Complex intensity and near-field description. *Report No. 46*, The Acoustics Laboratory, Technical University of Denmark, Lyngby, Denmark, 1991.
90. Maysenhölder, W., The reactive intensity of general time-harmonic structure-borne sound fields. In *Proceedings of the Fourth International Congress on Intensity*, ed. M Bockhoff, Centre Technique des Industries Mécaniques, Senlis, France, 1993, pp. 63–70.
91. Krasil'nikov, V. N., Effect of a thin elastic layer on the propagation of sound in a liquid half space. *Sov. Phys. Acoustics*, **6** (1960) 216–24.
92. Jacobsen, F., Sound power determination using the intensity technique: state of the art. In *Proceedings of the Third International Congress on Recent Developments in Air- and Structure-borne Sound and Vibration*, ed. M.J. Crocker, Auburn University Alabama, USA, 1994, pp. 1023–30.
93. Hübner, G. and Rieger, W., Schallintensitäts-messverfahren zur Schallleistungsbestimmung in der Praxis. *Schriftenreihe der Bundesanstalt für Arbeitsschutz*, Bundesanstalt für Arbeitsschutz, Dortmund, Federal Republic of Germany, 1988.
94. Jacobsen, F., Spatial sampling errors in sound power estimation based upon intensity. *J. Sound Vib.*, **145** (1991) 129–49.
95. Jacobsen, F., Random errors in sound power determination based upon intensity measurement. *J. Sound Vib.*, **131** (1989) 475–87.
96. Tachibana, H., Yano, H. and Yamaguchi, K., The accuracy of scanning sound intensity method in sound power level determination. In *Proceedings of Inter-Noise 93*, eds P. Chapelle and G. Vermeir. The Belgian Acoustical Association, BAV, The Technical Institute, KVIV, Belgium, Leuven, 1993, pp. 357–62.
97. Pettersen, O. K. Ø., and Newman, M., Is our confidence in scanned intensity measurements justified? In *Proceedings of Inter-Noise 89*, ed. George C. Maling, Noise Control Foundation, New York, 1989, pp. 979–84.
98. Olsen, H., Pettersen, O. K. Ø., and Vigran, T. E., Measuring strategies using the scanning intensity technique. In *Proceedings of the 1992 Nordic Acoustic Meeting*, ed. M. Newman, NTH, Trondheim, Norway, 1992, pp. 22–4.
99. Jacobsen, F., Sound power determination using the intensity technique in the presence of diffuse background noise. *J. Sound Vib.*, **159** (1992) 353–371.
100. Williams, R. G. D. and Yang, S. J., Sound field characterisation and implications for industrial sound intensity measurements. *Applied Acoustics*, **35** (1992) 311–23.
101. Pettersen, O. K. Ø and Newman, M.J., The determination of radiated sound power using sound intensity measurements. Report No STF44 A86166, ELAB, Elektronikklaboratoriet ved NTH, N-7034, Trondheim, Norway, 1986.
102. Tachibana, H. and Yano, H., Changes of sound power of reference sound sources influenced by boundary conditions measured by the sound intensity technique. In *Proceedings of Inter-Noise 93*, ed. P. Chapelle and G. Vermeir. The Belgian Acoustical Association, BAV, The Technical Institute, KVIV, Belgium, Leuven, 1993, pp. 1009–14.
103. Wei, W., Hickling, R. and Leek P., Gated sound-power measurement using an automated sound-intensity system. *Noise Control Engng Jour.*, **39** (1992) 13–19.
104. Degeorges, J. F., Rayonnement acoustique des plaques en champ proche. Thèse de Docteur, Université du Maine, Le Mans, France, 1988.
105. Adin Mann III, J. and Tichy, J., Near-field identification of vibration sources, resonant cavities, and diffraction using acoustic intensity measurments. *J. Acoust. Soc. Amer.*, **90** (1991) 720–9.
106. Reinhart, T. E. and Crocker, M. J., Source identification of a diesel engine using acoustic intensity measurements. *Noise Control Engng*, **18** (1992) 84–92.

107. Johns, W. D. and Porter, R.H., Ranking of compressor station noise sources using sound intensity techniques. *Noise and Vibration Control*, **19** (1988) 70–5.
108. Pepin, H., Localisation des sources de bruit sur une machine industrielle. In *Proceedings of the Second International Congress on Acoustic Intensity*, ed. M. Bockhoff. Centre Technique des Industries Mécaniques, Senlis, France, 1985, pp. 413–20.
109. Birembaut, Y., Deschamps, M. and Tourret, J., Noise generation study of synchronous belts using acoustic intensity. In *Proceedings of Inter-Noise 84*, ed. George C. Maling. Noise Control Foundation, New York, 1984, pp. 1155–60.
110. Kristiansen, U. R., Pure tone noise from idling circular saw blades. In *Proceedings of Inter-Noise 80*, ed. George C. Maling. Noise Control Foundation, New York, 1980, pp. 329–32.
111. Oswald, L. J., Identifying the noise mechanisms of a single element of a tire tread pattern. In *Proceedings of Inter-Noise 81*, eds L. H. Royster, D. Hunt and N. D. Stewart. Noise Control Foundation, New York, 1981, pp. 53–6.
112. Rasmussen, G. and Rasmussen, P., Localization of sound sources. In *Proceedings of the Second International Congress on Acoustic Intensity*, ed. M. Bockhoff. Centre Technique des Industries Mécaniques, Senlis, France, 1985, pp. 425–32.
113. Bucheger, D. J. and Trethewey, M. W., A selective two-microphone intensity method for evaluating sound radiation from individual elements of a coherent source array. In *Proceedings of Inter-Noise 82*, ed. James G. Seebold. Noise Control Foundation, New York, 1982, pp. 683–6.
114. Wagstaff, P. R. and Henrio, J. C., The measurement of acoustic intensity by selective techniques with a dual channel FFT analyser. *J. Sound Vib.*, **94** (1984) 156–9.
115. Bohineust, X., Wagstaff, P. R., Henrio, J. C. and Pasqualoni, J. P., Acoustic intensity vectors and sweep spatial averaging using the selective technique. In *Proceedings of Inter-Noise 87*, ed. Li Peizi. Acoustical Society of China, P.O. Box 2712, Beijing, People's Republic of China, 1987, pp. 1215–18.
116. Fuller, C. R., Structural influence of the cabin floor on sound transmission into aircraft—analytical investigations. *J. Aircraft*, **24** (1987) 731–6.
117. Lang, M. A., Sound intensity/power as a noise control diagnostic tool. In *Proceedings of Inter-Noise 89*, ed. George C. Maling, Noise Control Foundation, New York, 1989, pp. 973–8.
118. Oshino, Y. and Arai, T., Sound intensity in the near field of sources. In *Proceedings of the Symposium on Acoustic Intensity*, Tokyo, 1987, pp. 46–56.
119. Clark, A. R. and Watkinson, P. S., Measurements of underwater acoustic intensity in the nearfield of a point excited periodically ribbed cylinder. In *Shipboard Acoustics*, ed. J. Buiten. Kluwer Academic Publishers, Dordrecht, The Netherlands, 1986, pp. 177–88.
120. Watkinson, P. S., Measurements of sound absorption using sound intensity techniques. In *Proceedings of the Autumn Conference 1982*, Institute of Acoustics, Edinburgh, 1982, pp. B.4.1–B.4.4.
121. Fahy, F. J., Rapid method for the measurement of sample impedance in a standing wave tube. *J Sound Vib.*, **97** (1984) 168–70.
122. Elliott, S. J., A sample two-microphone method of measuring absorption coefficient. *Acoustics Letters*, **5** (1981) 39–44.
123. Lahti, T., Application of the intensity technique to the measurement of impedance, absorption and transmission. In *Proceedings of the Second International Congress on Acoustic Intensity*, ed. M. Bockhoff. Centre Technique des Industries Mécaniques, Senlis, France, 1985, pp. 519–26.
124. Hewlett, D. A. K., The acoustical properties of 'Diagon' sound absorbers. BSc Dissertation, Institute of Sound and Vibration Research, University of Southampton, 1986.

125. Allard, J. F. and Sieben, B., Measurements of acoustic impedance in a free field with two microphones and a spectrum analyser. *J. Acoust. Soc. Amer.*, **77** (1985) 1617–18.

126. Allard, J. F., Bourdier, R. and Bruneau, A. M., The measurement of acoustic impedance at oblique incidence with two microphones. *J. Sound Vib.*, **101** (1985) 130–2.

127. Sato, T., Ishikawa, M., Wada, S. and Koyasu, M., Measurement of sound absorption coefficient in an office using sound intensity. In *Proceedings of Inter-Noise 81*, ed. George C. Maling, Noise Control Foundation, New York, 1981, pp. 1221–5.

128. Migneron, J.-G. and Asselineau, M., Utilisation de l'intensimétrie pour l'analyse du comportement d'une façade soumise à l'impact du bruit. In *Proceedings of the Second International Congress on Acoustic Intensity*, ed. M. Bockhoff. Centre Technique des Industries Mécaniques, Senlis, France, 1985, pp. 477–84.

129. Mason, J. M. and Fahy, F. J., The use of acoustically tuned resonators to improve the sound transmission loss of double panel partitions. *J. Sound. Vib.*, **124** (1988) 367–79.

130. Crocker, M. J., Raju, P. K. and Forssen, B., Measurement of transmission loss of panels by the direct determination of transmitted acoustic intensity. *Noise Control Engng*, **17** (1981) 6–11.

131. Cops, A. and Minten, M., Comparative study between the sound intensity method and the conventional two-room method to calculate sound transmission loss of wall constructions. *Noise Control Engng*, **2** (1984) 104–11.

132. Van Zyl, B. G. and Erasmus, P. J., Applications of sound intensimetry to the determination of sound reduction indices in the presence of flanking transmission. In *Proceedings of the Second International Congress on Acoustic Intensity*, ed. M. Bockhoff. Centre Technique des Industries Mécaniques, Senlis, France, 1985, pp. 555–60.

133. Tachibana, H., Applications of sound intensity technique to architectural acoustic measurement (in Japanese). In *Proceedings of the Symposium on Acoustic Intensity*, Tokyo, 1987, pp. 103–14.

134. Guy, R. W. and De Mey, A., Measurement of sound transmission loss by sound intensity. *Canad. Acoust.*, **13** (1985) 25–44.

135. Halliwell, R. E. and Warnock, A. C. C., Sound transmission loss: comparison of conventional techniques with sound intensity techniques. *J. Acoust. Soc. Amer.*, **77** (1985) 2094–103.

136. Nielsen, T. G., Intensity measurements in builing acoustics. Application note, Bruel and Kjaer, Naerum, Denmark, 1986.

137. Van Zyl, B. G., Erasmus, P. J. and Anderson, F., On the formulation of the intensity method for determining sound reduction indices. *Appl. Acoust.*, **22** (1987) 213–28.

138. Jonasson, H. G., Sound intensity and sound reduction index. *Applied Acoustics*, **40** (1993) 281–93.

139. Olsen, H. and Newman, M.J., Determination of sound reduction indices using intensity techniques in situ. *Report No. STF40 A92045*, Acoustics Research Center, SINTEF DELAB, N-7034, Trondheim, Norway, 1992.

140. Puts, B. H. C. M. and Cauberg, J. J. M., Determination of transmission loss of saddle roof constructions. In *Proceedings of the Second International Congress on Acoustic Intensity*, ed. M. Bockhoff. Centre Technique des Industries Mécaniques, Senlis, France, 1985, pp. 503–10.

141. McGary, M. C. and Mayes, W. H., A new method for separating airborne and structure-borne aircraft interior noise. *Noise Control Engng*, **20** (1983) 21–30.

142. Hee, J., Gade, S. and Ginn, K. B., Sound intensity measurements inside aircraft. Application note, Bruel and Kjaer, Naerum, Denmark.

143. Pascal J. C., Practical measurement of radiation efficiency using the two microphone method. In *Proceedings of Inter-Noise 84*, ed. George C. Maling. Noise Control Foundation, New York, 1984, pp. 1115–20.

144. Steyer, G. C., Alternative spectral formulations for acoustic velocity measurement. *J. Acoust. Soc. Amer.*, **81** (1987) 1955–61.

145. Forssen, B. and Crocker, M. J., Estimates of acoustic velocity, surface velocity and radiation efficiency by using the two-microphone technique. *J. Acoust. Soc. Amer.*, **73** (1983) 1042–53.

146. Loyau, T., Determination of radiation efficiency with a three microphone probe. In *Proceedings of Inter-Noise 90*, ed. H. G. Jonasson, Acoustical Society of Sweden, Göteborg, Sweden, pp. 1053–6.

147. Watkinson, P. S., Personal communication.

148. Alfredson, R. J., The direct measurement of acoustic energy radiated by a punch press. In *Proceedings of Internoise 80*, ed. George C. Maling. Noise Control Foundation, New York, 1980, pp. 1059–66.

149. Nelson, P. A. and Elliott, S. J., *Active Control of Sound*. Academic Press. London, 1992.

150. Bullmore, A. J., Elliott, S. J. and Nelson, P. A., Mechanics of active suppression of harmonic enclosed sound fields. In *Proceedings of Euromech Symposium 213, Méthodes actives de contrôle du bruit et des vibrations*, ed. B. Nayroles. CNRS, Laboratoire de Mécanique et d'Acoustique, Marseille, France, 1986.

151. Loyau, T., Pascal, J. C. and Gaillard, P., Broadband acoustic holography reconstruction from acoustic intensity measurements. I: Principle of the method. *J. Acoust. Soc. Amer.*, **84** (1988) 1744–50.

152. Adin Mann III, J. and Pascal, J. C., Locating noise sources on an industrial air compressor using broadband acoustical holography from intensity measurements (BAHIM). *Noise Control Engng Jour.*, **39** (1992) 3–12.

153. Laville, F., Sidki, M. and Nicolas, J., Spherical acoustical holography using sound intensity measurements: Theory and simulation. *Acustica*, **76** (1992) 193–8.

154. Pierce, A. D., *Acoustics: an Introduction to its Physical Principles and Applications*. McGraw-Hill, New York, 1981.

155. Mason, J. M. and Fahy, F. J., Application of a reciprocity technique for the determination of the contributions of various regions of a vibrating body to sound pressure at a receiving point. *Proceedings of the Institute of Acoustics*, **12** (Part 1) (1990) 469–76.

156. Holland, K. R. and Fahy, F. J., A simple volume velocity transducer. In *Proceedings of Inter-Noise 93*, eds P. Chapelle and G. Vermeir. The Belgian Acoustical Association, BAV, The Technical Institute, KVIV, Belgium, Leuven, 1993, pp. 1261–4.

157. Mason, J. M. and Fahy, F. J., Measurement of the sound transmission characteristics of model aircraft fuselages using the reciprocity technique. *Noise Control Engng Jour.*, **37** (1991) 19–29.

158. Verheij, J. W., Reciprocity method for the quantification of airborne sound transfer from machinery. In *Proceedings of the Second International Congress on Recent Developments in Air- and Structure-borne Sound and Vibration*, ed. M.J. Crocker, Auburn University, Alabama, USA, 1992, pp. 591–8.

159. Zheng, J., Fahy, F. J., and Anderton, D., Application of a vibro-acoustic reciprocity technique to the prediction of sound radiated by a motored I.C. engine. *Applied Acoustics*, **42** (1994) 333–46.

160. Guy, R. W. and Abdou, A., A measurement system and method to investigate the directional characteristics of sound fields in enclosures. *Noise Control Engng Jour.*, **42** (1994) 8–18.

161. Tachibana, H., Visualisation of sound fields by the sound intensity technique (in Japanese). In *Proceedings of the Second Symposium on Acoustic Intensity*, Tokyo, 1987, pp. 117–26.

162. Tro, J., Sound radiation from a double bass visualised by intensity vectors. *Report No. STF44 A83088*, ELAB, Elektronikklaboratoriet ved NTH, N-7034 Trondheim-NTH, Norway, 1983.

163. Maynard, J. D. and Williams, E. G., Calibration and application of nearfield holography for intensity measurement. In *Proceedings of Inter-Noise 82*, ed. James G. Seedbold. Noise Control Foundation, New York, 1980, pp. 707–10.

164. Strøm, S., Ottesen, G., Pettersen, O. K. Ø and Johannesen, A., Concert hall acoustics—measuring the sound intensity in a simulated stage opening with and without diffusing elements in the stage. *Report No. STF44 A84001*, ELAB, Electronikklaboratoriet ved NTH, N-7034 Trondheim-NTH, Norway, 1984.

165. Kaiser, J. M. and Leeuwestein, A., Calculation of the sound radiation of a non-rigid loudspeaker diaphragm using the finite-element method. *J. Audio Engng Soc.*, **36** (1988) 539–51.

166. Munjal, M. L., *Acoustics of Ducts and Mufflers*. John Wiley, New York, 1987.

167. Terao, M. and Sekime, H., On acoustic energy flow around a rectangular square elbow. In *Summaries of Technical Papers of the Architectural Institute of Japan* (1984), pp. 99–100.

168. Fahy, F. J., Sound intensity distribution in ducts and branched pipes. In *Proceedings of the Second International Congress on Acoustic Intensity*, ed. M. Bockhoff. Centre Technique des Industries Mécaniques, Senlis, France, 1985, pp. 177–83.

169. Smith, J., Sound power flow in ducts carrying airflow. MSc. Dissertation, Institute of Sound and Vibration Research, University of Southampton, 1987.

170. Fahy, F. J., Sound power distribution in branched piping systems. *Proc. Inst. Acoust.* **7** (1985) 281–9.

171. Coulson, R., An experimental study of the dissipation of sound energy in a plenum chamber. BSc Dissertation, ISVR, University of Southampton, 1987.

172. Morfey, C. L., Acoustic energy in non-uniform flow. *J. Sound Vib.*, **14** (1971) 159–70.

173. Jacobsen, F., Measurement of sound intensity in the presence of airflow. In *Proceedings of the Second International Congress on Acoustic Intensity*, ed. M. Bockhoff. Centre Technique des Industries Mécaniques, Senlis, France, 1985, pp. 193–200.

174. Munro, D. H. and Ingard, K. U., On acoustic intensity measurements in the presence of mean flow. *J. Acoust. Soc. Amer.*, **65** (1979) 1402–16.

175. Chamant, M., Conception d'une sonde destinée à la mesure de l'intensité acoustique dans un fluide en mouvement. Thèse présentée devant l'Université Claude-Bernard, Lyon, France, 1981.

176. Seybert, A. F., Two-sensor methods for the measurement of sound intensity and acoustic properties in ducts. *J. Acoust. Soc. Amer.*, **83** (1988) 2233–9.

177. Chung, J. Y. and Blaser, D. A., Transfer function method of measuring acoustic intensity in a duct system with flow. *J. Acoust. Soc. Amer.*, **68** (1980) 1570–7.

178. Vasudevan, M. S. and Fahy, F. J., Comment on 'Transfer function method of measuring acoustic intensity in a duct system with flow'. *J. Acoust. Soc. Amer.*, **79** (1986) 1180(L).

179. Davies, P. O. A. L., Bhattacharya, M. and Bento-Coelho, J. L., Measurement of plane wave acoustic fields in flow ducts. *J. Sound Vib.*, **72** (1980) 539–42.

180. Åbom, M. and Bodén, H., Error analysis of two-microphone measurements in ducts with flow. *J. Acoust. Soc. Amer.*, **83** (1988) 2429–39.

181. Fahy, F. J., Lahti, T. and Joseph, P., Some measurements of sound intensity in

airflow. In *Proceedings of the Second International Congress on Acoustic Intensity*, ed. M. Bockhoff. Centre Technique des Industries Mécaniques, Senlis, France, 1985, pp. 185–92.

182. Fahy, F. J., Measurement of sound intensity in low-speed turbulent airflow. In *Proceedings of Inter-Noise 88*, ed. M. Bockhoff. Centre Technique des Industries Mécaniques, Senlis, France, pp. 83–7.

183. Broch, M., Wind and turbulence noise of turbulence screen, nose cone and sound intensity probe with windscreen. In *Bruel and Kjaer Technical Review No 4— 1986*, Bruel and Kjaer, Naerum, Denmark, 1986, pp. 32–9.

184. Fuller, C.R. and Fahy, F.J., Characteristics of wave propagation and energy distribution in cylindrical elastic shells filled with fluid. *J. Sound Vib.*, **81** (1981) 501–18.

185. Badie-Cassagnet, M., Bockhoff, M. and Lambert, J.M., Application de l'intensimétrie acoustique à l'identification des sources de pulsation de pression dans des circuits. In *Proceedings of the First International Congress on Acoustics*, ed. M. Bockhoff. Centre Technique des Industries Mécaniques, Senlis, France, 1981, pp. 253–60.

186. Verheij, J.W., On the measurement of energy flow along liquid-filled pipes. In *Proceedings of the Sound International Congress on Acoustics*, ed. M. Bockhoff. Centre Technique des Industries Mécaniques, Senlis, France, 1985, pp. 201–8.

Index